"[Mr. Aczel] does what the best writers on difficult subjects do: He takes his reader by the hand and explicates matters by returning to basics, in this instance what exactly it means when we count numbers. . . . Highly enjoyable."

—*The New York Times*

"[A] fascinating . . . introduction to an amazing and sometimes baffling set of problems, suited to readers interested in math—even, or especially, if they lack training."

—*Publishers Weekly*

"[A] well-written, witty book . . . even nonmathematicians will be carried along by a narrative with the pace of a thriller."

—*The New Scientist*

"[I]ndispensable."

—*Booklist* (starred review)

"[A]n engaging . . . explanation of the mathematical understanding of infinity, enlivened by a historical gloss on the age-old affinities between religious and secular conceptions of the infinite."

—*The Washington Post*

**Available in paperback
from Washington Square Press**

Léon Foucault (1819–1868)
Self-portrait, one of the earliest daguerreotypes, 1840s. *(CNAM, Paris)*

PENDULUM

*Léon Foucault and
the Triumph of Science*

AMIR D. ACZEL

WASHINGTON SQUARE PRESS
NEW YORK   LONDON   TORONTO   SYDNEY

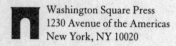 Washington Square Press
1230 Avenue of the Americas
New York, NY 10020

ISBN: 0-7434-6478-8
      0-7434-6479-6 (Pbk)

First Washington Square Press trade paperback edition September 2004

10 9 8 7 6 5 4 3 2 1

Washington Square Press and colophon are
registered trademarks of Simon & Schuster, Inc.

Manufactured in the United States of America

For information regarding special discounts for bulk purchases,
please contact Simon & Schuster Special Sales at 1-800-456-6798
or business@simonandschuster.com

For Miriam, who found the first *Arago* in Paris

CONTENTS

—◆—

CONTENTS

"The phenomenon develops calmly, but it is inevitable, unstoppable. One feels, one sees it born and grow steadily; and it is not in one's power to either hasten it or slow it down. Any person, brought into the presence of this fact, stops for a few moments and remains pensive and silent; and then generally leaves, carrying with him forever a sharper, keener sense of our incessant motion through space."

—Léon Foucault,
describing his pendulum
experiment, 1851

PREFACE

———◆———

The main events described in this book took place in the course of one year, 1851. They happened in Paris, or more precisely, at the intellectual center of the French capital. In fact, the locations of the three main events of this book form the three corners of a perfect imaginary triangle lying in the heart of the Left Bank of Paris. This triangle encompasses within it the elegant Luxembourg Garden, the Latin Quarter with its universities and cafés, and the fashionable district below the ancient church of Saint Germain des Prés—the areas of Paris that a few decades later would become the favorite haunts of writers and artists. But in the mid-1800s, scientific history was made here, and our understanding of the universe changed forever. The change was brought on by the work of one man, a lone Parisian genius who was neither a trained scientist nor educated at the famous

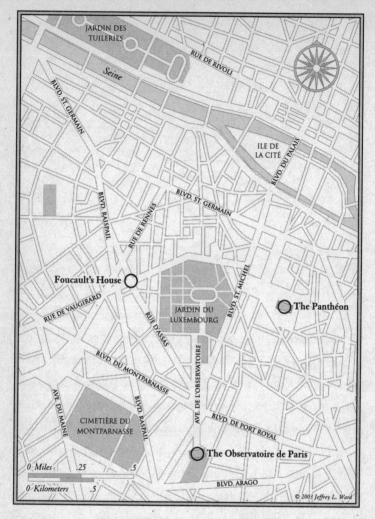

Paris and the three points of the triangle.

universities that had made the French capital the leading center of ideas and learning. This is the story of Jean Bernard Léon Foucault (1819–1868); and of his pendulum, with which he showed us that the world turns, putting an end to centuries of persistent skepticism and conflict between science and faith.

Foucault was much more than the inventor of the pendulum experiment. While not trained in science, he had an incomparable ability to understand nature, as well as a legendary dexterity. These skills allowed him to carry out the demanding pendulum experiment, to build new telescopes, invent regulators for stage lighting, improve photographic techniques, measure the speed of light in air and in water, and invent the gyroscope.

But despite his great achievements, recognition came slowly to Foucault. The scientific establishment did not want to accept him. He was not a member of the club, as it were, and the mathematicians thought he had no mathematical ability and hence could not possibly address the problems of physics in any meaningful way. And yet Foucault was able to go beyond designing and performing experiments: Without being a mathematician, he developed the mathematical law governing the rate at which his pendulum moved away from its original plane of oscillation as a function of the latitude at which it was located—a discovery that shocked and embarrassed the mathematicians. While French mathematicians and physicists refused to recognize his genius, foreign organizations credited Foucault's achievements long before he was acknowledged for them in his native country. He was awarded Britain's coveted Copley Medal in 1855, a decade before receiving comparable honors in France.

In France, it took a decree by an emperor, Napoléon III, to give Foucault the accolades he deserved. Napoléon also made Foucault the Physicist Attached to the Imperial Observatory in Paris, forcing the exclusive Parisian science establishment to accept the man and his achievements. Napoléon III ensured that Foucault's discoveries and inventions be remembered, by commissioning a publication of his life's work. This is the story of the unusual partnership between emperor and unappreciated genius, the story of a pendulum that taught us that the world turns, and a tale of the triumph of science over ignorance.

1

———◆———

A STUNNING DISCOVERY
IN THE CELLAR

From his journal, we know that he made the discovery at exactly two o'clock in the morning on January 6, 1851. He was down in the cellar of the house he shared with his mother, located at the corner of the rue de Vaugirard and rue d'Assas—in the heart of the intellectual Left Bank of Paris and within the immediate area in which Gertrude Stein and Picasso would live during the next century. He had been working feverishly in the cellar for weeks, but no one walking on the fashionable street above could suspect that down below an experiment was being prepared— one that would forever change the way we view the world.

Jean Bernard Léon Foucault (Léon Foucault to all who knew him) was thirty-two years old. He was not a trained scientist, but he already had a few scientific achievements to his credit, including a clever experiment to measure the speed of light. And he could claim credit for some inventions as well, including a design for light in microscopy and a way of regulating theatrical lighting. But during the last few months of 1850 and into 1851, Léon Foucault had been concentrating all his efforts on a different kind of problem. He was attempting to solve the most persistent scientific problem of all time: one that had plagued Copernicus, Kepler, Descartes, Galileo, and Newton in the sixteenth to the eighteenth centuries, and that—surprisingly—remained unresolved as late as Foucault's own time.

He had prepared his experiment carefully, perfecting it during long hours of concentrated work in his cellar over a period of months. Foucault's remaining problems with the experiment were technical ones, and he was an expert at doing precision work with his hands. He worked with wires, metal cutters, measuring devices, and weights. He finally secured one end of a 2-meter long steel wire to the ceiling of the cellar in a way that allowed it to rotate freely without resulting torque. At the other end of the wire, he attached a 5-kilogram bob made of brass. Foucault had thus created a free-swinging pendulum, suspended from the ceiling.

Once the pendulum was set in motion, the plane in which it oscillated back and forth could change in any direction. Designing a mechanism that would secure this property was the hardest part of his preparations. And the pendulum had to be perfectly symmetric: Any imperfection in its shape or distribution of weight could skew the results of the experiment, denying Foucault the proof he desired. Finally, the pendulum's swing had to be initiated in such a way that it would not favor any particular direction because a hand pushed it slightly in one direction or another. The initial conditions of the pendulum's motion had to be perfectly controlled.

Since such a pendulum had never been made before, the process of building it also required much trial and error, and Foucault had been experimenting with the mechanism for a month. Finally, he got it right. His pendulum could swing in any direction without hindrance.

On January 3, 1851, Foucault's apparatus was ready, and he set the device in motion. He held his breath as the pendulum began to swing. Suddenly the wire snapped, and the bob fell heavily to the ground. Three days later, he was ready to try again. He carefully set the pendulum in motion and waited. The bob swung slowly in front of his eyes, and Foucault attentively followed every oscillation.

Finally, he saw it. He detected the slight but clearly perceptible change he was looking for in the plane of the swing of the pendulum. The pendulum's plane of oscillation had moved away from its initial position, as if a magic hand had intervened and pushed it slowly but steadily away from him. Foucault knew he had just observed the impossible. The mathematicians—and among them France's greatest names: Laplace, Cauchy, and Poisson—had all said that such motion could not occur or, if it did, could never be detected. Yet he, not a mathematician and not a trained physicist, somehow always knew that the mysterious force would be there. And now, he finally found it. He saw a clear shift in the plane of the swing of the pendulum. Léon Foucault had just seen the Earth turn.

2

ANCIENT LOGIC:
BIBLE AND INQUISITION

Foucault knew the importance of his discovery. His clear and simple proof of the rotation of the Earth would have far-reaching implications for society, culture, and the relationship between religion and science. He was well aware of the long and ago-nized history of the problem he was addressing with a pendulum swinging in the cold, damp cellar that January night in 1851.

Two and a half centuries earlier, on February 19, 1600, the Inqui-sition brought the Italian monk and teacher Giordano Bruno (1548–1600) in chains to Campo dei Fiori, in the center of Rome, tied him to an iron stake, and burned him alive. One of Bruno's crimes was his belief that the Earth rotated.

A third of a century later, Galileo was put on trial in Rome by the same Inquisition. Threatened with torture, humiliated, forced to kneel before his prosecutors, the great scientist who had discovered the moons of Jupiter, sighted the rings of Saturn, and explained to us much about the physical world was made to recant his belief that the Earth turned. Only this move would save him from a painful death—the fate of Giordano Bruno—and allow his sentence to be commuted to house arrest for the remainder of his life. But the ordeal broke his spirit and damaged his health, and he died a few years later.

The Inquisition's reign of terror continued through the centuries, with the burning of books whose content deviated from strict Church dogma, the listing of banned books, and the prosecution of anyone who promulgated views that differed from those the Church believed were in agreement with scripture.

Just what were these views? They were inspired by biblical passages. In the Book of Joshua we read:

> Then spake Joshua to the Lord in the day when the Lord delivered up the Amorites before the children of Israel, and he said in the sight of Israel, Sun, stand thou still upon Gibeon; and thou, Moon, in the valley of Ajalon. And the Sun stood still, and the Moon stayed, until the people had avenged themselves upon their enemies. Is not this written in the book of Jashar? So the Sun stood still in the midst of heaven, and hasted not to go down about a whole day.[1]

And the Book of Isaiah contains the passage:

And this shall be a sign unto thee from the Lord, that the Lord will do this thing that he hath spoken. Behold, I will bring again the shadow of the degrees, which is gone down in the sundial of Ahaz, ten degrees backward. So the Sun turned ten degrees, by which degrees it was gone down.[2]

Ecclesiastes includes the well-known sentence: "The Sun also ariseth, and the Sun goeth down, and hasteth to his place where he ariseth again."[3]

The Roman Catholic Church held that the Copernican view that the Earth turns—rather than the Sun—was clearly at odds with these biblical references. The Church, through the Inquisition, was determined to stamp out any contentions that the Earth turned, labeling such views as heretical. And the consequences of professing such beliefs were clear to anyone living in the lands in which the Church exerted its influence.

The question of whether the Sun and Moon orbit a stationary Earth or whether the Earth rotates and orbits the Sun has its origins long before the time of the Inquisition. And while the Hebrew Bible contains simple descriptions of risings and settings of the Sun and the Moon, this doesn't mean that peoples of antiquity uniformly believed in a stationary Earth. Some Greek philosophers indeed held that the Earth stood motionless under rotating heavens, but others believed that the world turns and that its rotation on its axis gives us the illusion of the risings and settings of stars, Sun, and Moon.

Plato and Aristotle (fourth century, B.C.) clung firmly to the belief that the Earth is immobile and that the firmament with its stars and planets, as well as Sun and Moon, rotates around the Earth. Aristotle's philosophy gained support in medieval and later times, and the Church adopted it for its use.

Another fourth-century, B.C., Greek philosopher, Philolaus of Crotona, professed the opposite view. Philolaus was a member of the Pythagorean school, established in Crotona, in southern Italy, in the previous century by Pythagoras. In his *Treatise of the Sky,* written sometime in the middle of the fourth century, Philolaus wrote: "Of an opposite opinion are the representatives of the Italian School called the Pythagoreans. For them, it is the fire that occupies the center; the Earth is only one of the moving stars, and its circular motion about its own center produces the day and the night. They also construct another Earth, opposite of ours, which they call the anti-Earth."[4]

While there was a complication in this ancient view of the universe—the existence of an anti-Earth—the theory seems surprisingly accurate: The Sun is in the center of the solar system; the Earth and the other planets orbit the Sun; and the Earth rotates about its axis, producing the day and the night. Two other fourth-century Greek thinkers, Heraclides and Nicetas, also believed in a rotating Earth, as did Aristarchus of Samos, who lived a century later. These philosophers realized that the simplest way to explain the apparent movements of the stars, planets, Sun, and Moon was to assume that the Earth itself moved. Since they believed that the simplest theory to explain nature was probably the right one, these ancient scholars embraced the heliocentric view of the universe, in which the Earth is one of the planets, rotating about its axis and orbiting the Sun.

The heliocentric theory is simpler for several reasons. Look at the night sky for several hours, and you will notice that *all* the stars move uniformly from east to west. There are two ways to explain this. First, the stars—somehow—all have exactly the same speed as they travel through the night sky overhead. The speeds must all be the same, or else the relative shapes of the constellations would change as one star overtook another, and we know that this never happens (at least not over a single night of observation). The other possibility is that we, the observers, together with our Earth, rotate in space in the opposite direction to that of the apparent motion of the stars. The situation is similar to that of a person looking out the passenger window of a moving car: Do the trees all move away from the car at exactly the same speed, or is it the passenger in the car who is moving away from all the stationary trees?

Clearly, the hypothesis that the Earth rotates is much simpler than the hypothesis that the Earth is stationary and the stars all move away at a uniform rate. Another reason why the moving-Earth theory is simpler has to do with *retrograde* motion of planets. This backward-movement of a planet is detected sometimes when the position of a planet is measured over several nights of observation. The phenomenon occurs when Earth "overtakes" a planet in their common race around the Sun (as seen from our vantage point on Earth). The simpler way to explain this curious effect is to assume that Earth rotates around its axis and orbits the Sun, and that so does the planet in question. A comparison of the two orbits can then reveal when the planet should appear to an observer on Earth to move away from its expected course in the sky over a period of time. To explain retrograde motion in another way, one that maintains a stationary Earth, is difficult.

So already in antiquity, keen observers of the sky, mathematicians and philosophers, developed a system for the solar system that was correct in its essence and had the Earth rotating about its axis and orbiting the Sun.

Unfortunately, history would drown these ancient voices of truth because others would argue more forcefully for a *complicated,* rather than simple, theory of the universe—one that could explain all the astronomical observations while still maintaining the special role that people seemed to want the Earth to enjoy, that of center of all Creation.

It thus happened that Aristotle's theory that the Earth was the stationary center of the universe found an unlikely proponent half a millennium later, in Alexandria, Egypt, in the person of the greatest astronomer of the ancient world: Claudius Ptolemy (second century, A.D.).

Ptolemy gathered all the knowledge of astronomy available by his time and published it in a book called the *Almagest.* The title was a Latin adaptation of an earlier, Arabic name of the work (itself derived from the original Greek) meaning "The Greatest." The book consisted of thirteen volumes and included a compilation of centuries of astronomical observations, as well as discussions of trigonometry used in astronomical analysis. It also included Ptolemy's model of the universe.

This gifted Alexandrian astronomer constructed an ingenious theory of the world, the stars, the planets, the Sun, and the Moon. In Ptolemy's model all the motions of the stars and planets were explained by a complicated series of circles, and circles-within-circles. Some of these circles were epicycles. An epicycle is a circle whose center lies on the circumference of a larger circle. As the larger circle moves around its center, so does the smaller circle. Thus a point on the circumference of the smaller circle exhibits the complex motion that results from moving about a center that, itself, rotates about another center. This compound movement described well the retrograde motion of planets. The model had a circle describing the celestial sphere containing all the stars, as well as circles and epicycles for the planets and the

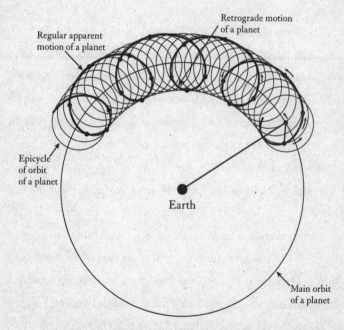

How the use of an epicycle explains retrograde motion of a planet.

Moon, and a circle for the Sun. At the center of the entire model was a stationary Earth. Ptolemy's mathematical genius thus allowed people to maintain the belief in a geocentric universe, while explaining all the apparent motions of the heavenly bodies. No wonder his model survived (although challenged) to Foucault's time—seventeen centuries.

The Church taught the ideas of Aristotle about the perfection of nature, justifying them by using Ptolemy's model of the universe, which agreed with the Church's own interpretation of the statements in the Bible. The Church used the Ptolemaic theory to attack the opposing Copernican view of the universe, which emerged in the sixteenth and seventeenth centuries.

Nicholas Copernicus (1473–1543) was born on February 19, 1473, in Toruń, Poland, to a wealthy family of merchants. (While universally known by the Latin form of his name, the boy was born Mikolaj Kopernik.) Copernicus studied mathematics and astronomy for four years at the University of Krakow, which was an important center for both disciplines, boasting two chaired professorships in astronomy.

He later moved to Italy and began the study of canon and civil law at the University of Bologna—one of the oldest universities in the world. But his love of astronomy was rekindled when he made the acquaintance of a professor of astronomy at Bologna, Domenico Maria Novara, from whom he rented a room. Copernicus began to work with Professor Novara, helping him record observations of the sky. On March 9, 1497, the two witnessed an occultation of the star Aldebaran by the Moon. (An occultation takes place when one celestial body hides another. Here, the Moon, which was close to Aldebaran, slowly moved to cover the bright star, a rare occurrence.) This astronomical event impressed the student, and he later mentioned it in his writings as one of the experiences that motivated his nascent theory. He also took sightings of the Sun, using a new instrument he invented, which allowed him to reflect an image of the Sun on a wall. But most of the knowledge Copernicus gathered about astronomy came from reading books. And one of the books he read most carefully was Ptolemy's *Almagest*. Copernicus admired Ptolemy's genius as evidenced by the writings of the ancient astronomer.

But Copernicus's careful study of the book revealed some errors. Whenever Copernicus tried to use the model of the universe presented in the *Almagest* to predict astronomical events, he found that these predictions were off by hours or days. For example, the length of the year, as predicted by the Ptolemaic system, was not what he knew it should be. Copernicus soon realized that Ptolemy's entire underlying model had to be wrong, and he set out to correct it. The result was Copernicus's own book on astronomy, called *De Revolutionibus Orbium Coelestium* ("On the Revolutions of the Celestial Spheres"), in which he advocated the view that the Sun, not the Earth, was at the center of the solar system. In coming to his conclusions, Copernicus may have been influenced by the work of the ancients, since he mentioned Aristarchus of Samos in his writings. Copernicus placed the planets in their orbits around the Sun in what we know today is the correct order of increasing distance from the Sun: Mercury, Venus, Earth, Mars, Jupiter, and Saturn.

As soon as he devised his model, Copernicus found that it worked better than the old Ptolemaic system. Predictions of astronomical events were much more accurate than those produced by the old model. One example of the accuracy of the Copernican model was its prediction of the size of the Moon. In Ptolemy's model, the Moon's motion was explained by an epicycle; but the epicycle also implied a significant variation in the size of the Moon (because its distance from Earth changed much). In reality, and according to the theory of Copernicus, this variation is small.

Copernicus escaped the Inquisition—he died before the fury could be unleashed on him. *De Revolutionibus* was published the year he died, 1543. According to tradition, Copernicus received the first copy of his book on his deathbed, and died holding it in his hand.

Once the theory that the Sun was at the center of the solar system and the Earth rotated on its axis and orbited the Sun like the other planets was published and shown to be accurate, people began to pay attention. These people were the more educated and wealthy from among Europe's populations—those who could read and afford to buy books. The printing press had been invented several decades earlier, and books became more freely available about the time of the appearance of *De Revolutionibus*. Scientists studied the Copernican theory and, despite objections from the Church, attempted to publish opinions and results in agreement with the new theory.

But, as earlier in history, there were steps backward. Tycho Brahe (1546–1601), an illustrious Danish astronomer who lost his nose in a duel and thereafter wore a nose made of gold, compiled a voluminous list of astronomical observations. His records were of unprecedented accuracy—ten times more precise than any observations made before his time. Brahe considered his astronomical data so important, that he kept his observations secret.

Brahe had built an observatory on an island given him for that purpose by the Danish king. When the royal successor was uninterested in supporting his projects, Brahe moved to Prague and established a new base of astronomical work there. On November 11, 1572, Brahe observed a rare astronomical event: a supernova—something we know today is the explosion of a dying star. (In our time, only one supernova in our galaxy has been seen, a 1987 explosion observed from Earth's southern hemisphere.) A supernova appears like a new, bright star in the sky (hence "nova," for new). This observation contradicted Aristotle's belief in the immutability of the heavens. But Brahe was not prepared to confront the Inquisition, and in the model of the universe he derived using his exquisite set of observations, he still placed an immobile Earth in the center of the universe. In Brahe's model of the solar system, the Sun and Moon circled the Earth, while the five known planets circled the Sun.

In his waning years, Brahe was forced to take on an assistant, a young German mathematical genius by the name of Johannes Kepler (1571–1630). At first, the suspicious Brahe refused to share his wealth of astronomical data with his new assistant, but these observations inevitably came into Kepler's hands.

The use Kepler made of Brahe's astronomical observations is considered one of the most amazing achievements in the history of science. Kepler analyzed Brahe's data over a period of years, long after he had inherited his position as astronomer following Brahe's death. When he was done, he had in front of him the natural law governing the movement of planets.

Kepler's laws are still used today by NASA to determine the precise locations of planets to which spacecraft are to be sent; and astronomers have used Kepler's laws to determine the orbits of planets outside our solar system, which in the 1990s started to be discovered.

Briefly, Kepler's laws state that planets move in ellipses, and that the Sun (or another star, in the case of extra-solar planets) is at one of the foci of the ellipse. A planet sweeps in its orbit equal areas in equal time. Thus it is the *area* of the ellipse mapped out by the planet's orbit that is preserved as constant in time, and not the distance traveled by the planet. This was a surprising and counterintuitive finding.

In the following century, Isaac Newton (1642–1727) would spend years to complete a proof that his own overarching law of universal gravitation implied Kepler's laws as a special consequence. The fact that Kepler had derived these laws without the advantage of Newton's (and Leibniz's) calculus, developed in the next century for this kind of analysis, is a testament to his mathematical ability.

Kepler believed in the Copernican view of the universe,

rather than in his mentor Brahe's model. His work later served to update the Copernican system, since in Copernicus's model—as well as in all previous models of our solar system—the orbits of the planets are circles, while Kepler showed them to be ellipses.[5]

Kepler's brush with the authorities, however, was not due to his Copernican views. Rather, it was because his mother was accused of being a witch, which was not a rare charge against a woman in Europe in those days. But the scientist was able to testify on her behalf, and she was acquitted. Later, he turned his mathematical ability to the problem of estimating the volumes of wine casks. This was especially useful in 1612, a very good year for wine.

The great problem faced by anyone who tried to argue for the Copernican system was that no one possessed a *clear proof* that the Earth rotated.

There were astronomical observations that could be interpreted as supporting the moving-Earth hypothesis, but—equally—one could use these same observations in other ways, as Ptolemy had done in antiquity and as Brahe had done at the end of the sixteenth century.

The Church had its own astronomers (and there is an active Vatican Observatory today), and some of them were very good. In the 1600s, the Church astronomers argued for those models of the universe that had a nonmoving Earth in the center of the system.

One of the Church astronomers was P. Christoph Scheiner (1573–1650). Scheiner used a telescope designed by Kepler, which he adapted for observing an image of the Sun reflected on a wooden board. When he studied the Sun in this way, he found sunspots. The Society of Jesus, of which Scheiner was a member, decided that his observations could not be valid. With Aristotle, the Church believed in the perfection of the Sun, and the sunspots jeopardized this belief. Scheiner thus faced opposition from his own peers and difficulties in disseminating his results. So Scheiner should have been somewhat favorably inclined toward another scientist who had also observed sunspots, the great Galileo Galilei (1564–1642).

But this was not to be. Galileo, who had a certain arrogance about his scientific and mathematical abilities, antagonized Scheiner by ridiculing his interpretation of sunspots. (Scheiner thought they were little satellites obstructing our view of the Sun.) Scheiner never forgot the insult and years later joined his detractors when Galileo was under attack by the Church for his support of the Copernican theory.

More than any other tragedy of the Inquisition, the story of Galileo underscores the lack of a convincing proof of the rotation of the Earth. In 1609, Galileo began to build telescopes, having heard that such an instrument had just been invented by the Dutch. He pointed his telescopes toward the sky and made discoveries of immense importance. On January 7, 1610, while observing Jupiter, Galileo noticed three little "stars," arranged in a line around Jupiter: two to the left, and one to the right. The next day, he saw the three "stars" all to the right of Jupiter; and on January 10, he saw only two "stars," both to the left of the planet. Galileo concluded that the "stars" were satellites, like our Moon, in orbit around Jupiter. Today we call them the Galilean satellites.

Galileo attached great importance to his discovery, since he believed it provided evidence that Earth was certainly not the center of *everything* in the universe. For here were three heavenly bodies (a fourth was seen shortly afterwards; and a dozen smaller ones have been discovered in modern times) in orbit around a center other than our Earth. Galileo, who had always been a staunch supporter of the Copernican theory, became much bolder in his statements in support of the heliocentric model. Thus he incurred the wrath of the Inquisition. Galileo had no *definitive* proof—other than astronomical observations—that the Earth itself rotated. And thus his insistence that Copernicus was right, his overconfident attacks on the Church, and his naïve belief that if he came to Rome to present his case to the Church he would prevail—all without incontrovertible proof of his claims—only hastened his downfall.

As he watched his majestic pendulum swinging back and forth that night in the cellar two centuries after Galileo's death, Foucault could not have missed the irony, for it was Galileo himself who first studied the pendulum.

As a young man, Galileo once attended a service in the great Cathedral of Pisa. It was a windy day; the windows of the church were wide open, and the large chandelier overhead was swinging in the wind. Curious about the phenomenon of a swinging object, Galileo timed the movements of the chandelier against his pulse. He discovered that long and short swings all took about the same time to complete. This, Galileo's first-discovered physical law, is known as the law of the *isochronism* of the pendulum.[6]

Had Galileo spent many more hours observing a pendulum swing in a controlled environment (and had he known what to look for and prepared the experiment accordingly), perhaps he might have noticed the slow drifting of the plane of oscillation of the pendulum—a proof that the Earth is rotating under the pendulum. This was the very phenomenon Foucault had just observed in his cellar.

And there is some evidence that the Church might well have accepted such evidence, ending 2,000 years of the Aristotelian-Ptolemaic doctrine. We know that on April 12, 1615, Cardinal Bellarmine, who had been involved in the Inquisitorial process against Giordano Bruno and who was to become instrumental in the prosecution of Galileo, wrote to Paolo Antonio Foscarini (1588?–1616) that if a proof could be given that the Earth rotated, the Church would change its view.[7] As we will see, that would indeed happen in time. Had Galileo possessed such irrefutable proof—of the kind Foucault now had—he perhaps would have been saved from the terrible ordeal that ultimately cost him his life and profoundly affected the behavior of the scientists who followed him over the decades and centuries to come.

• • •

There is a bizarre footnote to this story. In an obscure paper of 1660 or 1661, cited in a manuscript of 1841 found at the library of the Grand Duke of Tuscany, Vincenzo Viviani (1622–1703), who was a student of Galileo, wrote cryptically: "We observe that all the pendulums hanging on one thread deviate from their initial vertical plane, and always in the same direction." Viviani did not elaborate, and the work itself disappeared. It was rediscovered only after Foucault's pendulum experiments became public in 1851.[8]

3

---◆---

FAILED EXPERIMENTS WITH FALLING BODIES

On a late spring day in 1638, passersby in the French countryside outside Paris witnessed a most unusual sight. A monk dressed in robes and a hood, and his assistant wearing a uniform with a cape and a bonnet, stood next to a cannon aimed straight up into the sky—in a perfectly vertical direction. The monk held a long staff, the end of which was aflame, and he touched the flame to the primer of the cannon. The cannon fired, sending a cannonball straight upward. The two men remained standing next to the cannon, looking up. But nothing happened next: not a minute later, not ten minutes later, not even an hour later. The cannonball had mysteriously disappeared into the sky and was never found again—fortunately for the two men, who remained standing bewildered next to their cannon. This event has been

Marin Mersenne's cannon experiment inspired by Descartes.
(Bibliothèque nationale de France)

called "the most stupid experiment" ever associated with the name of Descartes.[9]

The monk who performed this bizarre test was Marin Mersenne (1588–1648), a close friend and confidant of the French mathematician, physicist, and philosopher René Descartes (1596–1650). The monk's assistant was the engineer Pierre Petit, who was the superintendent of fortifications at the military base from which the two men borrowed the cannon.

Descartes had suggested an experiment with a cannon because he was desperate for proof of the rotation of the Earth. Descartes was a staunch supporter of the Copernican view of the universe, and he had hoped that a cannonball might fall down to Earth in a way that could prove that the Earth rotated. He never expected that the cannonball would simply disappear—but the experiment itself was the result of a sequence of misunderstandings.

Since cannons first began to be used in warfare, artillery officers had noticed a consistent deviation of cannonballs away from their targets. In particular, a cannonball fired northward from a cannon located anywhere in the northern hemisphere will deviate east of the location at which it is expected to fall. The reason for this variance is the Coriolis effect (discussed in Chapter 9), which also explains why water generally drains in a counterclockwise rotation in the northern hemisphere and clockwise south of the equator.

Today we recognize René Descartes as one of the greatest scientists and philosophers of all time. He was born on March 31, 1596, in La Haye, near Tours, France, to a noble family of modest means. Descartes was educated at the Jesuit College at La Flèche, and upon graduation entered the University of Poitiers to study law. He soon lost interest in the law and left for Paris, where he spent his time gambling. Later he became a professional soldier and took part in a number of military campaigns in Holland and Bohemia. In military encampments over years of fighting, Descartes began to develop his scientific ideas—among them his masterpiece: the mathematical field of analytic geometry. He became familiar with cannons and used trajectories of cannonballs and bullets to study gravity. In the spring of 1620, Descartes saw heavy fighting in the battle of Prague and almost died. The result of this experience was his internal conflict between religious feelings and the call of science. He was painfully aware of a potential contradiction between the two and was especially distraught by the Church's vehement rejection of the ideas of Copernicus. Descartes's own theories in physics were closely allied to the belief that all objects in our solar system orbited the Sun. He felt stifled, constrained, held back from letting the world know about his great scientific discoveries.

Fearing persecution for his ideas, which were clearly against Church teachings, Descartes spent the next twenty years wandering through Holland, a country in which liberal views were welcome and in which the Catholic Church had little influence. The Holy See had condemned the Copernican theory in Italy, France, Germany, and Belgium in an edict of 1616. This left Holland as the only country in that part of Europe in which Descartes felt he could be safe.

Ironically, his closest ally throughout this period was a Catholic priest—Marin Mersenne. Father Mersenne had attended the same Jesuit college at La Flèche, graduating some years before his younger friend. The priest developed a love of science and was open-minded about theories of nature. His interests ranged from physics to number theory.[10] Throughout Descartes's travels in Holland, during which he frequently changed his address and worried about being pursued by agents of the Church because of his heretical ideas, he maintained an active correspondence with Mersenne, discussing mathematical and philosophical ideas.

Much of what we know about Descartes comes from the letters he wrote to Mersenne, because Mersenne was a meticulous keeper of records and upon his death left 10,000 letters from Descartes and other scientists and mathematicians. Many of these letters have been published in France, and excerpts of a few of them are given below. They shed light on the life of the scientist-philosopher and on his fear of the Inquisition.[11]

On February 1, 1634, Descartes wrote Mersenne from Deventer.

My Reverend Father,

. . . The knowledge I have of your virtues makes me hope that you will still have a good opinion of me once you see that I have voluntarily and entirely canceled the treatise, thus losing four years of my work, in order to give full obeisance to the Church, since it defends its opinion about the movement of the Earth. And anyway, since I have seen neither the Pope nor the Council ratifying such a defense, done only by the Congregation of Cardinals established for the censorship of books, I know well that now in France their authority is such that they can make of it an article of faith. I allow myself to say that the Jesuits have aided in the condemnation of Galileo, and all the books of P. Scheiner provide enough proof that they are not his friends. But otherwise the observations in that book furnish enough support for the movements he attributes to the Sun, that I am of the belief that P. Scheiner himself is not of the opinion of Copernicus. I find this so astonishing that I do not dare publish my own sentiment. For me, I search for nothing but rest and the tranquility of spirit, goals that cannot be achieved by those given to animosity. I wish only to instruct others, especially those that have already acquired some credit for false opinions, and who have some fear of loss lest the truth be revealed.

I am your very obedient and very affectionate servant,

Descartes.

The letter gives us a view of the tormented scientist. He had spent four years writing a scholarly book and had promised to send it to Mersenne. Then, for fear of a fate similar to Galileo's, he retracted the book and confessed it to his friend, who happened to belong to the same Church that continued to persecute scientists and free thinkers who did not espouse its views about the universe.

In another letter from Deventer, at the end of February 1634, he wrote:

> . . . No doubt you know that Galileo had been convicted not long ago by the Inquisition, and that his opinion on the movement of the Earth had been condemned as heresy. Now I will tell you that all the things I explain in my treatise, among which is also that same opinion about the movement of the Earth, all depend on one another, and all are based on certain evident truths. Nevertheless, I will not for the world stand up against the authority of the Church . . . I have the desire to live in peace and to continue on the road on which I have started.

Descartes's next letter, from Amsterdam on August 14, 1634, contained the following:

> I begin to feel sorrow for not having any news from you. . . . Mr. Beeckman came here on Saturday evening and brought me the book of Galileo, but took it with him to Dort this morning, so I had the book in my hands for only 30 hours. I haven't had the opportunity to read it all, but I can see that he philosophizes well about the movement of the Earth, even though it is not enough to be convincing. . . . For what he says about a cannon fired parallel to the horizon, I believe that you will find quite a sensible difference, if you should perform this experiment yourself.

This was the beginning, the seed of the idea that ultimately resulted in Mersenne finding himself in the middle of a field in the French countryside firing a cannon straight up into the sky. It started with Descartes's obsession with finding a proof of the rotation of the Earth—perhaps as an effort to vindicate Galileo. But clearly Mersenne misunderstood how to design the experiment. For the cannon should have been fired "parallel to the horizon" and not straight up into the sky.

While he had been careful not to offend the Church, Descartes's scientific writings were still seen to support the Copernican theory. His books contained technical physical descriptions of cannonballs and musket bullets fired in different directions, and how they were affected by the rotation of the Earth. Eventually, these writings were condemned as heretical, and Descartes's books were listed on the Church's *Index of Prohibited Books*. But Mersenne—the cleric—continued to publish Descartes's work in Paris, in defiance of his own Church. Eventually, Dutch Protestant theologians also condemned Descartes's writings, describing them as "atheistic," and his standing in Holland became precarious. Reluctantly, Descartes began to look for another home.

The opportunity came in 1646, when Queen Christina of Sweden invited Descartes to join her court as the Royal Philosopher. Descartes was ambivalent—he was still attached to his life in Holland—but the queen persisted, preying on his admiration and respect for royalty. Finally, in the spring of 1649, the queen sent Admiral Fleming of the Swedish Royal Fleet on a special mission to pick up the philosopher and bring him to Sweden.

The queen required Descartes to give her daily lessons in philosophy at 5 A.M. in the unheated library in her palace. His new lifestyle quickly took its toll, and early in 1650, Descartes fell ill. He died on February 11. Two decades later, the French Government brought his remains back to France and reburied them in the Panthéon in Paris.

Descartes never found a proof of the rotation of the Earth using cannonballs or bullets, but his line of thought was to guide others in this direction.

When Isaac Newton made his famous statement "If I have been able to see farther, it was only because I stood on the shoulders of giants," he meant Descartes, Kepler, and Galileo.[12] Isaac Newton was born in Woolsthorpe, England, on Christmas Day in 1642—the year in which Galileo died—and his life's work can be viewed as a direct continuation of the noble attempts by these three scientists to understand the universe in which we live and to decipher its laws.

Newton was a country boy; he grew up on a farm in Lincolnshire. But his teachers recognized early on that the boy had a superior intellect and encouraged him to apply to Cambridge University. Newton enrolled at Trinity College, Cambridge, in 1661, and eight years later he was appointed the Lucasian Professor of Mathematics in Cambridge (this chair is held today by Stephen Hawking).

The reason for this recognition was Newton's immense achievements in mathematics and physics during the two years 1664–65. Bubonic plague arrived in England in 1664, and the universities were closed. Newton left Cambridge and returned to Woolsthorpe, where he spent the next two years meditating and writing. It was here in the countryside of his youth that the apple fell on his head—a story true or apocryphal—turning his attention to the mystery of gravity. As a result of his deep study, Newton developed the calculus to enable him to derive the law of universal gravitation and his laws of motion. That same year, he also proved that white light is composed of all the colors of the rainbow.

Newton's laws confirmed in his mind that the Earth rotated on its axis and orbited the Sun. Kepler's laws all made perfect sense now, in view of the larger theory of gravitation, and the Earth had to follow the same laws obeyed by the other planets. Newton derived the law of motion of the Moon as well and explained many other phenomena based on the laws of gravity. Mathematics—the new calculus Newton had developed, which had been simultaneously discovered by Leibniz in Germany—made it possible to derive the laws of physics. But despite all the marvelous advances in mathematics and physics, most of them thanks to his own work, Newton was well aware of the lack of a definitive terrestrial proof of the rotation of the Earth.

The falling apple that may have led him to his great discoveries could be seen as a symbol of the law of gravity. And gravity was the force making celestial bodies rotate around one another (or, more precisely, about their common center of mass). Using astronomical observations, Newton knew that the Moon keeps "falling down to Earth" at a rate of 0.0045 feet per second by going around it. The Moon thus falls toward the Earth, but keeps "missing" us, and the falling-down motion becomes a "going around" motion. The Moon is like the mythological Sisyphus, forever rolling a rock up a steep hill, never making actual progress toward the destination, because the Moon's motion of falling down is transformed into orbital motion. Newton also knew that the Moon was at a distance of roughly sixty times the radius of the Earth away from us. He thus used the *inverse-square* law of gravity, which he had derived, to conclude that an apple in Lincolnshire, which is one Earth-radius away from the center of the Earth, should therefore fall down to Earth (in its first second of fall) at a speed of sixty *squared* times the rate at which the Moon "falls" towards us. This gave him the number 60 x 60 x 0.0045 = 16.2 feet, which is very close to the actual value of the acceleration due to the force of gravity on Earth.[13] The connection, therefore, between orbital motion of heavenly bodies and the effects of gravity we experience in everyday life on Earth is direct and tangible, as Newton understood.

The Sun, being so much larger than the Earth and the other planets, had the same effect on these bodies, making them forever "fall" toward it by an eternal rotation around it. In 1679, Newton came to the conclusion that a falling object, such as an apple, might provide terrestrial evidence for the rotation of the Earth.

Newton put this idea in a letter of November 28, 1679, to the Royal Society in London. He suggested that if falling objects were to be studied, a consistent deviation east would be found, due to the rotation of our planet. Thus, studying how falling objects behaved, he argued, should provide the needed proof of the rotation of the Earth. He wrote: "If a ball should fall every time east of the bar, this would constitute a proof of the diurnal rotation of the Earth." Newton never carried out such an experiment, but Robert Hooke (1635–1703), who was the secretary of the Royal Society, followed Newton's suggestion and did perform such tests shortly afterwards. On January 6, 1680, Hooke wrote to Newton:

> In the mean time I must acquaint you that I have (with as much care as I could) made 3 tryalls of the experiment of the falling body, in every of which the ball fell towards the south-east of the perpendicular, and that very considerably, the last being above a quarter of an inch, but because they were not all the same I know not which was true.[14]

Hooke's experiments with falling objects thus revealed not only the correct deviation east, but also a smaller deviation to the south, now known to be spurious. These results were seen as inconclusive.

Here, too, the ghost of Galileo hovers. Hooke's experiments, and the ones to follow, reflected studies of falling objects carried out a few decades earlier by Galileo. Galileo was concerned with the problem of the time it takes heavy and light objects to fall to the ground from a given height. And he found that light objects and heavy objects took the same time to fall down, if one ignored the resistance of the air (which affects larger objects more than smaller ones). Galileo is reputed to have carried out these experiments from the top of the Leaning Tower of Pisa, and we can visualize the old master leaning over the edge of the tower, which overlooks this magnificent city, methodically dropping lead weights to the ground.

And once again, Galileo came tantalizingly close to obtaining evidence for his claim that the Earth rotated. For the objects Galileo dropped to the ground had to have deviated east of the vertical—but he was completely unaware of it.

Newton came to his conclusion that falling objects should deviate east of their expected landing points because he understood nature incredibly well. The laws he had formulated, Newton's laws of motion, told him how falling objects should behave. Newton explained the principle as follows. A ball is held in the hand of a person standing on the top of a tower. The Earth's rotation gives the top of the tower a given radial velocity—the speed at which the top of the tower travels around through space. That velocity is higher than that of a point down below the tower, because that point is closer to the center of the Earth, and the center of the Earth has no velocity at all as the Earth turns. When the person on top of the tower drops the ball, it begins to fall down to Earth. As the ball falls down, it maintains the horizontal speed it has acquired by being on top of the tower.

(The reason for this is one of Newton's laws: A body will maintain its inertia—here the continuation of its movement in the horizontal direction—unless acted on by a force; and there is no force here to stop this motion.) Thus the ball progresses east (along the horizontal, east is the direction of the movement due to the rotation of the Earth) more than does a point at the bottom of the tower. And therefore the ball will hit the ground east of that point. We also know that this phenomenon is a manifestation of the force of Coriolis (explained in Chapter 9), although Newton could not have called it by that name—Coriolis was yet to be born.

Experiments with falling bodies continued, and a century later, Abbé Giambattista Guglielmini (1764–1817) in Bologna dropped objects from the Torre degli Asinelli, 241 feet high. He found an average deviation of 18.89 millimeters east, consistent with a rotating Earth, but again a slight southern deviation confounded the results.

In 1802, Johann Friedrich Benzenberg (1777–1846) studied Guglielmini's results and in July and October of that year carried out his own experiments of falling bodies from a 235-foot tower in Hamburg. He obtained an average deviation of 9 millimeters east and 3.5 millimeters south. Benzenberg consulted the German astronomer Heinrich Olbers (1758–1840), who in turn discussed the problem with the famous mathematician Carl Friedrich Gauss (1777–1855), then only twenty-five. Gauss and, independently of him, the French mathematician Pierre-Simon de Laplace (1749–1827) developed theories for and studied the results of falling objects in an effort to prove that the Earth spins about its own axis. Laplace believed that it was illogical to assume that the Earth was immobile. In his 1796 treatise, *L'exposition du système du monde,* he had written:

> Isn't it infinitely more simple to suppose that the globe we inhabit rotates about itself than to assume that a mass as considerable and as distant as the Sun moves with the extreme speed necessary for it to turn around the Earth in one day?[15]

Laplace believed that scientists had a responsibility to find a proof of the rotation of the Earth, and at the beginning of the nineteenth century he wrote:

> The rotation of the Earth must be established with all the certitude that can be provided by the physical sciences. A direct proof of this phenomenon should be of interest to geometers and physicists alike.[16]

Both Gauss and Laplace found no southern deviation in their computations of the expected results of experiments. Gauss computed the results Benzenberg should have obtained, and found that these were 3.95 millimeters east and 0 millimeters south, rather than what Benzenberg actually obtained. These experiments continued well into the nineteenth century. In 1831, Ferdinand Reich (1799–1882) conducted experiments with falling objects in a 158.5-meter deep mine shaft near Freiburg,

Germany. Despite many precautions, Reich's results were not better than those of Benzenberg.[17]

But all the experimental data collected over the years did not amount to a proof of the rotation of the Earth. The results for falling objects were never clear-cut, and there was always some other deviation from the expected landing point, so that no one could determine that the deviation was definitely east, as the theory required. Winds, temperature variations, measurement errors, and other factors may have muddled the results. Science had no dramatic verification of the rotation of the Earth, and the world did not overthrow the Church-supported Aristotelian view that we remain motionless in a universe rotating about us. While many educated people in Europe and elsewhere believed that Earth rotated about its axis and revolved around the Sun, the world was still in need of a definitive, terrestrial proof.

Tantalizingly, astronomical evidence that accorded with the hypothesis of a moving Earth mounted during the eighteenth century. But while such developments were welcomed by people who followed scientific advances and who already believed that the Earth turns and orbits the Sun, astronomical observations were never enough to convince either the Church or the common people.

Newton had argued in his famous book, the *Philosophiae Naturalis Principia Mathematica* (1687), that the force of gravitational attraction worked everywhere in our solar system. In 1758, Halley's Comet returned, and an astronomical study of its behavior provided more evidence that Newton was right. For here, a visitor to our solar system reacted to the gravitational attraction of the Sun, in its movement in a very long and eccentric orbit around it. In 1767, the Cambridge astronomer John Michell (1724–1793) proved that the phenomenon of the existence of double stars, observed by astronomers and amateurs everywhere in the sky, could not be due to chance. He showed that most of these double stars were binary pairs orbiting each other (rather than, by coincidence, appearing in the same place in the sky while actually located far apart). He argued that Newton's laws of gravitation were thus at play throughout the universe. If the laws of gravitation made objects in the universe orbit one another, would Earth be an exception to the cosmic rule?

In 1729, the astronomer James Bradley (1693–1762) found an annual aberration in the positions of fixed stars, which could only be explained scientifically by a rotating Earth in orbit about the Sun. The observations involved parallax: a triangulation using measured angles to distant stars. These angles were slightly different in the summer from their values observed in winter, and this aberration had to be due to the fact that Earth was located at different positions along its course around the Sun at different times of the year. Today, parallax is regularly used for estimating the distances to stars. The famous mathematician Jean Le Rond d'Alembert (1717–1783) later called Bradley's finding "the greatest discovery of the 18th century." But all such astronomical evidence was not enough. It did not constitute incontrovertible proof of the rotation of the Earth.

• • •

Despite Copernicus, despite Kepler's laws of planetary motion, despite Galileo's discovery of the moons of Jupiter, despite the substantial mathematical and scientific work of Newton, despite the astronomical observations in agreement with a moving Earth, despite all the advances in science, it was now 1851 and the world still had no clear, terrestrial proof that the Earth rotated. Not until now. Finally, a decisive proof was in front of Foucault in that cold cellar at 2 A.M. on January 6. But how would he convince the world that he had the ultimate proof of the rotation of the Earth? Foucault was not trained as a scientist. He was a dabbler in science, at most an amateur in the eyes of the academics of his day. How could this untrained lover of science convince the world that he had the final answer to humanity's quest? All we know is that the careful observer picked up his notebook and wrote, in his precise, rounded handwriting:

2 o'clock in the morning, the pendulum has moved in the direction of the diurnal motion of the celestial sphere.[18]

4

———◆———

A SCIENCE "IRREGULAR" IN THE AGE OF THE ENGINEER

Why was it Foucault? Why not Galileo, two centuries earlier? Why not Newton, with his supreme intellect and overarching theories that captured so well the behavior of all objects under the pull of gravity? Why not any of a number of people, mathematicians and physicists who had studied nature over the previous decades and centuries? To answer this question, we need to look both at Foucault the person and at the time during which he lived.

Foucault possessed a mix of characteristics that enabled him to successfully tackle the problem of the rotation of the Earth despite the odds against him. Foucault was very self-confident. And he maintained this confidence despite a lack of formal education, despite an admitted dearth of mathematical ability, and despite opposition from his peers. His deep sense of confidence tended to irritate both friend and foe alike, but it provided him with the drive and the determination to pursue his goals. And despite a complete lack of formal preparation, Foucault had an innate ability to understand nature—an inborn feel for the physical world around him. This intuition told him what to look for, how to prepare for it, and how to detect it.

Finally, Foucault had a gift: He had a great ability to work with his hands—to design and build complicated instruments with great precision and care. As such, Foucault was a product of his time—the nineteenth century. The problem of constructing a pendulum that could swing in any direction was not trivial. It required a degree of skill in engineering. A regular pendulum would not do here. A scientist would have to think about how to design a system that would allow a pendulum that starts swinging in a given direction to change that direction without hindrance, and in a way that no particular direction would be favored by an imperfection in the device. Knowing how to do this was probably not natural to a pure scientist such as Newton or Galileo or Kepler or Descartes. For the most part, these great minds were observers of nature. And while Newton and Galileo did design telescopes, for example, engineering skills were not common during their time, and they were primarily scholars rather than builders. But in the nineteenth century, engineering grew to become an important occupation. And Foucault was a quintessential engineer, in addition to being a keen observer of nature. He had witnessed and participated in the technological and scientific revolution that swept over the developed world in the nineteenth century. The 1800s were the heyday of science and technology. In particular, the period from 1820 to 1880 has been called "The Age of the Engineer."

Building on the earlier achievements of the late eighteenth century and the Industrial Revolution, enterprising scientists and engineers in Britain and France, as well as in the United States, set out to put these seeds of progress to use. It would be difficult to describe all the marvelous inventions and discoveries of the age. The most important ones were those related to the use of steam.

In 1761, the Scottish engineer James Watt (1736–1819) began his groundbreaking experiments in the production of power through the generation of steam, in continuation of the pioneering work of the English inventor Thomas Newcomen (1663–1729), who as early as 1712 had built a rudimentary steam-powered machine. In 1765 Watt built a condenser engine, and seven years later, a rotative steam engine. This design was very important in practical applications of the use of steam as a tremendous source of power in industry and transportation.

The steam engine became commonplace in the 1800s. The age of the steamship began in 1812, when small sailing ships were first equipped with steam engines. A rich British hotelier, Henry Bell, was the first to use a steamboat. The *Comet* ferried passengers along the Clyde River to Bell's hotel from 1812 until 1820. In 1815, William Hedley (1779–1843) built a steam-powered locomotive, inaugurating the era of trains. In 1829, Robert Stephenson (1803–1850) built an advanced, powerful steam locomotive called the "Rocket," and the world was now ready for the tremendous expansion in railway use and the resulting growth of trade and transportation.

Steel was invented in the 1850s, when the Englishman Henry Bessemer (1813–1898) conducted experiments to improve the strength and durability of iron for use in gun barrels. The invention of steel made machine production flourish, and large engines and turbines and machine tools of various kinds were invented and produced during this period. The Ames Gunstock lathe and other lathes essential in production processes were invented beginning in 1857. Marc Brunel (1769–1849), who fled the French Revolution to America and later settled in England, first started making block-making machines in 1803. These machines were used to make parts for other machines. In 1851, the first oil works opened, in Bathgate, Scotland. British Petroleum would later take charge of oil production from shale in this location, signaling the dawn of our motorized era worldwide.

In 1831, the English physicist Michael Faraday (1791–1867) set out the principle of electromagnetic induction, which quickly led to the generation of electric power from mechanical power, resulting in the invention of dynamos and electric power generators of all kinds, using steam and coal and oil, as well as the use of electricity in motors. Electric power had just been born, with all its implications for progress and change.

In 1837, William Cooke (1806–1879) and Charles Wheatstone (1802–1875) in the U.K. invented a practical telegraph system, which used five needles. Each needle, when used, would point either left or right. Any one of 20 letters of the alphabet could be transmitted using two needles. Thus there were 2 x 2 x 5 = 20 possible letters in the Cooke and Wheatstone alphabet, and the letters C, J, Q, U, X, and Z were left out. This made it difficult to transmit words like "quick" or "quiz," or "jinx." So for such words, alternative spelling was used. Still, the system worked well because it was intrinsically simple to use. In 1837, Cooke and Wheatstone made the first demonstration of its use between two railway stations in England, and the age of world-wide communication began. Twenty-one years later, in 1858, the first transatlantic cable was laid on the bottom of the ocean, connecting England with the United States.

Millard Fillmore was President of the United States at the time of Foucault's experiments in France. And, like Europe, the United States saw a tremendous technological and industrial growth during this time. German and Irish immigrants were flowing into the United States, bringing with them new ideas and great energy. The Irish immigration began in 1851, following the great potato famine. During the course of this migration, lasting until 1860, over 900,000 Irish immigrants would enter the United States. The U.S. population was 23 million in 1850, with fifteen percent of it living in cities.

Industrial production was expanding, and it was an exciting period in the life of the nation. The improvements in timekeeping brought about the factory bell, which regulated production activity. In 1845, Allen B. Wilson (1824–1888) invented the sewing machine, which *Scientific American* called "one of the greatest inventions of all time."

As happened in Europe, the United States also saw great social changes during this dynamic time of growth and expansion. Abolitionist writings proliferated, women fought for suffrage, and the new immigrants struggled to integrate into the culture of the New World. The United States saw a great expansion westward, with new technologies such as railroads, the telegraph, and improved transportation means helping people move to settle and industrialize the country.

Against this background of the whirlwind advances in technology that took place in quick succession in Europe and America, we see Léon Foucault, a man who loved machines and tools and who had worked with them as a young man. All of these new inventions and processes excited him and spurred him on: to invent, to discover, and to produce.

So who was Léon Foucault, the man who brought us the ultimate proof of the rotation of the Earth?

Jean Bernard Léon Foucault was born in Paris on September 18, 1819, to a comfortable middle-class family. His father, Jean Léon Fortuné Foucault, was a successful publisher who was known especially for his publication of a series of highly regarded volumes on the history of France. The father retired with his family to the city of Nantes, possibly because he suffered from poor health. He died there in 1829. After his father's death, his mother took the ten-year-old boy back to Paris. They settled in the comfortable house at the corner of the rue de Vaugirard and rue d'Assas. Léon Foucault would live with his mother in this house his entire life. He would never marry.

The house on the corner of the rue de Vaugirard and the rue d'Assas today. Note two commemorative reliefs: on the left, a pendulum; and on the right, a brief description of Foucault's life. *(Photograph courtesy of Debra Gross Aczel)*

The boy was weak and small, and he suffered from poor health. He was reserved, slow to respond, and reluctant to talk or act. Early biographers, who knew him personally, paint an unappealing portrait of a frail boy with a small head and asymmetrical eyes, which did not seem to both look in the same direction; one was myopic, and the other far-sighted. Foucault squinted often, and looked awkward. He was aware of these imperfections and therefore preferred to read alone rather than interact with people. His health problems would plague him throughout his life, worrying his mother, who would survive her son.

Mme. Foucault had great ambitions for her son. She sent him to the prestigious Collège Stanislas in Paris. But young Léon Foucault was not a good student, and for long periods of time he had to be educated at home by a tutor, since his school performance was so poor. He was habitually late with school assignments and seemed to his teachers to be lazy and timid. Jules-Antoine Lissajous (1822–1880), a childhood friend and member of the French Academy of Sciences, who wrote a memoir about him, presented to the Academy after Foucault's death, said: "Nothing about the boy announced that he would be illustrious some day; he was of delicate health, mild character, timid, and not expansive. The feebleness of his constitution and the slowness that characterized his work made it impossible for him to frequent a college. He thus could pursue his studies only thanks to the care of a devoted mistress under the watchful eyes of his mother."[19] Through the help of a series of tutors, Foucault was able to finish his years of schooling and earn his high school diploma, allowing him to continue his studies at a university.

As a boy, Foucault was clearly not interested in what he was taught at school. At the age of thirteen, he suddenly showed where his true talents, and his passions, lay. He began to work with his hands, using a variety of tools, working with great precision and care. He constructed various ingenious toys and machines. First, Foucault built himself a boat. Then he made a telegraph, which was a copy of the telegraph he observed in operation near his home, by the Church of Saint Sulpice. Foucault's little telegraph could emit a sequence of short and long beeps, which he used to transmit messages. We don't know which and how many letters Foucault's telegraph could transmit. Later, the boy constructed a small steam engine, which could really run. Foucault was evidently taken with the greatest inventions of his day and, even as a child, wanted to participate in the great revolution in engineering that was sweeping the Western world.

Foucault developed his early talent for working with his hands. Throughout his life, he worked with precision, and spent much time and effort designing useful devices and implements. The contraptions he built and inventions he made while still a boy impressed those around him more than did his shy, retiring personality. Some of the toys he made were preserved and were exhibited after his death.

His mother became convinced that the gift Foucault had of working with his hands would make him a successful surgeon. And since she wanted her son to become a medical doctor, she promptly enrolled him in medical school. Foucault entered medical school in Paris in 1839. He planned to specialize in surgery, as his mother had suggested. At medical school, he studied and participated in the hospital work required of all who aspired to become doctors. One of Foucault's favorite classes was a course of microscopy taught by Professor Alfred Donné (1801–1878). The admiration Foucault developed for Professor Donné seems to have been mutual. The professor recognized the talents of his young student and cultivated them. He entrusted his student with additional, special assignments and was impressed with his performance—Foucault seemed destined to a great career in medicine.

But soon something unforeseen happened. Foucault saw blood in the course of his work in the hospital, and it made him sick. In addition, the sight of the suffering of patients was too much for him to handle. He became so ill that he couldn't perform his duties any longer. Foucault had no choice: He had to withdraw from the medical program, to his mother's great disappointment.

But the bond Foucault had established with Professor Donné was strong, and the professor wanted the former student to continue to work with him. He recognized the young man's abilities, and his interest in medical science, and employed him as his assistant in a course of microscopy. If he couldn't become a medical professional, Foucault would work on the science of medicine.

While still a student of medicine, just before he dropped out, Foucault discovered the work of Daguerre on photography. Louis-Jacques Daguerre (1787–1851) invented daguerreotypy, an early method of photography, in 1837. Daguerre had begun his work as an apprentice in architecture, but later moved to work on stage design in the theaters of Paris. He invented a theater of illusion called *diorama*: a picture show with changing light effects and large stage-wide paintings used as background. Daguerre used a camera obscura as an aid in painting his theatrical background pictures. This led him to the idea of freezing the images, and he experimented with different methods of achieving this goal. One day in 1835, Daguerre by chance left an exposed plate in a cupboard in which there was a broken thermometer. The mercury fumes from the thermometer apparently reacted with the silver and other chemicals on the plate, and the image was set. Working on this process, Daguerre was able to develop his images in about thirty minutes.

On September 7, 1839, Daguerre, who had obtained significant government support for his photography projects, organized a series of public demonstrations of his photographic process at the Quai d'Orsay in Paris. The display of how photography works was such a success that Daguerre announced that France could now equip the entire world with the new apparatus for making photographs. He then turned the demonstrations of his new process into a free course to the public on photography, and the masses flocked to learn the new method of recording images. It was billed as a novel way of "painting" without the need for talent. In fact, daguerreotypy and the ensuing developments in photography were to affect the new style of painting that developed during this time: impressionism. Corot would take photographs of the scenes he wanted to paint back to his studio; Manet would base some of his portraits on photographs; and Monet owned four cameras and would base his famous painting *Women in the Garden* (1866) on a series of photographs.[20] Among the hundreds of people who flocked to Daguerre's course on photography was the twenty-year-old student of medicine, Léon Foucault.

Foucault had two good friends during this period—people

who knew him well and around whom he could be himself and not feel awkward because of his looks or his social reserve. One was his professor of microscopy, Alfred Donné, who would remain his friend and supporter throughout his life. The other was Foucault's old friend from the Collège Stanislas, Armand Hippolyte Louis Fizeau (1819–1896). Only a few days younger than Foucault, Fizeau came from an old family of doctors. His father was the chair of pathology at the medical school at which both young men were studying medicine. The boy was required to follow the family tradition and enroll in medical school. But like Foucault, Fizeau would drop out of medical school within a short time and pursue a career in science. Unlike his friend, however, he would do it formally by enrolling in a course in optics at the prestigious Collège de France, studying with a famous professor, Henri Victor Regnault (1810–1878).

Foucault shared his excitement about daguerreotypy with Fizeau, and the two friends decided to study Daguerre's process carefully. They came together to watch Daguerre present his method to the public. The master placed an apparatus by a window. He talked for a long time, while the apparatus—the camera—was still by the window. About half an hour later, he picked up the device and removed from it a copper plate covered with a thin coat of silver. Then he cleaned the plate carefully with oil of lavender. When it was perfectly clean, Daguerre evenly applied iodine to the plate and placed it inside a darkened chamber for several minutes. Daguerre picked up a beaker filled with mercury, heated it under a burner, and exposed the plate to the mercury vapor. Slowly, as if by magic, an image appeared on the plate. It was the exact negative of the view outside the window. Daguerre then washed the plate with salt water and showed it around the packed room to enthusiastic applause.

Foucault and Fizeau watched the process with great curiosity. They noticed that Daguerre always kept the camera still, for at least half an hour, before removing the plate and beginning the development process. And his subject was always the view outside the window, or a still object—never people. The two friends talked about it and decided that the technique might have much greater applicability if it could be used for making portraits. But people couldn't easily sit motionless for half an hour. And it seemed that Daguerre's process required that long to record the image.

Foucault and Fizeau began to experiment with daguerreotypy on their own. They experimented with various chemicals and—biographers say it was Fizeau who first had the idea—discovered that adding bromine to the plate sensitized it. Thus, in principle, using bromine to pretreat the copper plate coated with silver should allow for a shorter period of time to record the image. But how would they implement this discovery? The two young students worked for many days, and finally Foucault was able to determine the exact way of submitting the iodized plate to the action of the mercury vapor so that the necessary exposure time would be only twenty seconds. This would be their first joint scientific success. Unfortunately, neither of them was to share in the economic rewards of their discovery; for the most part, daguerreotypy was soon replaced by other methods of photography. But we do have a surviving handsome portrait of Foucault, a daguerreotype he made of himself during this period (see the frontispiece). The two friends enjoyed working together on the daguerreotypy project and would continue to collaborate.

After Foucault dropped out of the medical program, he worked with Professor Donné on microscopy. Donné's research involved the study of milk and other bodily fluids, as well as frogs' tongues. One of the problems that occupied them was that of providing adequate lighting for the microscope. Foucault, in particular, was now interested in using daguerreotypy to record images of objects seen through the microscope. He soon realized that the key to making high-quality pictures of objects under the microscope was having strong light, and that need lent urgency to his search for good lighting for microscopes. At that time, microscopes were lit with gaslight, which was uneven and weak. But electric power had recently been invented and everyone was excited about yet another invention with unlimited potential for improving life. Foucault turned his attention to electricity. He devised a carbon-arc electric lamp for the microscope. The arc had two separate parts with a point of contact, and Foucault noticed that as the carbon in the lamp burned, the intensity of the light changed because the distance between the two components kept changing due to the burning. He realized that the key to producing light of high intensity and constant output—crucial for daguerreotypy—was in regulating the distance between the two parts of the carbon arc. Foucault invented an electric regulator for this purpose by using electromagnets that constantly moved the two parts of the arc as the distance between them changed while the carbon burned.

The regulator ensured that the light output to the microscope was constant and its intensity sufficient for making pictures of the images under the microscope.

Foucault's invention of a light regulator for producing daguerreotypes of objects under the microscope found other uses as well. Among these applications was a regulator for stage lights. Daguerreotypes required natural light, and so Foucault's light source for making daguerreotypes of objects under the microscope was designed to produce light that was as close as possible to natural light. This kind of light is also very desirable on stage. Often a theatrical scene is staged to simulate the open-air in daytime. An electric arc by Foucault was used to simulate sunrise in Meyerbeer's opera *Le Prophète,* the first use of this technology for stage lighting. The invention was even presented to the French Academy of Sciences, but its young inventor did not receive any special recognition; a similar device had already been patented in England.

Foucault worked meticulously and carefully, leaving nothing to chance and always planning ahead down to the minutest detail. He earned tremendous respect from his professor, and Donné began to trust his young assistant more and more. In 1845, Donné's textbook, *A Course of Microscopy,* was published.[21] The second volume of the work was an atlas of eighty daguerreotypes of the subjects of microscopy discussed in the book, all made by Foucault, who was the coauthor of the volume. In the preface, Donné described his coauthor as: "a young scholar and distinguished amateur of photography."

• • •

Foucault continued to study light, photography, microscopes, and technical devices with enthusiasm. He made a few scientific advances, but since he was only a laboratory assistant to Donné and not a trained scientist, he did not earn much recognition.

Foucault did not have the mathematical background considered essential for understanding physics and did not hold a doctoral degree in science. He thus lacked the prerequisites for entrance into France's intellectual elite, a group so exclusive and aware of its social status that, within it, members addressed each other as "Cher Confrère Savant" (Dear Scholarly Colleague). Within this group, the mathematicians were the crème-de-la-crème. They were called "geometers," echoing the way mathematicians of ancient Greece referred to themselves, as they considered geometry the purest form of mathematics.

His lack of acceptance made Foucault bitter. He was a very gifted man, and he knew it. The confidence he had about his abilities worked against him. Throughout his life, many people viewed him as overly self-assured, and even arrogant. He was certainly quiet, but with an air of superiority, as some saw it. His attitude made the scientists trained at the vaunted French Grandes Écoles treat him dismissively.

In 1845, Alfred Donné retired from his duties as scientific editor for the newspaper *Journal des Débats* (the Journal of Debates), an important daily paper in France at that time. He passed the assignment on to his coworker, Léon Foucault. The science editor had the responsibility of reporting in the newspaper any important news about science discussed at the weekly Monday meeting of the Academy of Sciences. Foucault performed this job admirably and for years wrote occasional articles explaining science to the public. He wrote about trajectories of comets, total solar eclipses, and advances in chemistry. Through this job, which required attendance at the meetings of the Academy, Foucault got to know the Perpetual Secretary of the Academy of Sciences, as well as the many distinguished scientists who were members of this elite group of scholars.

Joseph Bertrand, who three decades later was the Perpetual Secretary of the Academy of Sciences, described Foucault in his short biography, published in 1882:

At the age of 25, having learned nothing in the schools, and even less from books, avid for science but loving study less, Léon Foucault accepted the mission of making known to the public the works of the savants and of judging their discoveries. From the beginning, he demonstrated much sense, much finesse, and a liberty of judgment tempered by more prudence than one would expect from a biting and severe spirit. His first articles were remarkable; they were spiritual. He took his task seriously. Introduced without apprenticeship and without a guide into this academic pell-mell, an abundant and confused mixture of all the problems and all the sciences, he showed no awkwardness, and, in a role in which mediocrity would never be tolerated, obtained a complete success.[22]

To compound his problems with the established academic community, through his articles in the *Journal des Débats*—which often expressed strong and controversial views about science—Foucault made some enemies and called unhelpful attention to himself. Bertrand described Foucault's role as a science journalist, revealing aspects of his personality, in the following words.

> Persons of considerable esteem in science solicited Foucault's attention, less fearful perhaps of his opinion than of his praise. Coolly polite, attentive only to the truth, Foucault judged with study and reflection, and without complacency. This unknown young man, who had no published scientific work to his name, no discovery justifying his quickly acquired authority, made them impatient with his assured tranquility, irritated them by his audacious frankness, exasperated them at times with his thin irony. . . . He excited lively resentments and gave rise to deep rancors.[23]

Bertrand gives an amusing example of the kind of article about science Foucault wrote for the *Journal des Débats*.

> At the last meeting of the Academy of Sciences, a very strange and very amusing invention was presented. This machine has, say its inventors, no furnaces, no boilers, no cylinders, no pistons, no wheels, and—we might add, for honesty's sake—no force! We make this declaration not

only for the sake of the author, but for the sake of the academician who has caused this dreadful racket by good-naturedly presenting us a steam tourniquet weighing ten kilograms as having the force of a horse![24]

Ultimately, Foucault was a science generalist. He was a good expositor of science to the average citizen, and was an excellent engineer, inventor, and self-made scientist. But he had not specialized in any particular area within science. He started in medical school, he became interested in photography, he worked as an assistant to a professor of microscopy, he was good in engineering, he invented a few things, but he wasn't trained in any of these fields, and he stuck to none of them. He had a love for a general subject called science. As another early biographer described him, Foucault was a "science irregular."

So how would this "science irregular" now announce to the world the earth-shattering discovery he had just made this night in January 1851? How would he convince the world of science, which had until now largely ignored his work and discoveries and whose experts had even gained some dislike for him, to listen? How would he convince the world that he had the ultimate proof of the rotation of the Earth? But if he could—if only he were allowed to show the scholars of his day that he had the final proof science so badly needed of Earth's rotation—the exclusive group of French science might finally also give him the credit he deserved for his previous achievements. Thus the pendulum he was now watching swing in the Parisian night might hold a hope of his redemption as a scientist and as a human being.

Late that fateful night in 1851, he closed his logbook and went upstairs to bed. What was on his mind? Could he sleep that night? Did he discuss his discovery with his mother? With a friend? At some point during the night, Foucault came to the conclusion that there was one person in the world who might help him: one man, a unique individual, whose life, background, status, and character could not have been more different from those of Foucault himself. This man was Foucault's only hope. But fate does work in mysterious ways.

5

THE MERIDIAN OF PARIS

Heading east from his house on the rue de Vaugirard (which happens to be the longest street in Paris) or walking south along the rue d'Assas, Foucault would shortly reach the magnificent Jardin du Luxembourg, a garden made for Princess Marie de Medici (1573–1642) when she missed the beautiful gardens of her native Florence after marrying King Henri IV and thus joining the French royal family in Paris in the early years of the seventeenth century. Today this beautiful garden rivals any in Italy. Foucault loved the Luxembourg Garden and spent many hours walking its lanes, contemplating problems of science. If Foucault were to leave the garden through its southern gate, he would find himself on the tree-lined Avenue de l'Observatoire, and at its end he would reach the Paris Observatory, a large building whose high white dome rules over the Paris skyline.

The Observatoire de Paris is the oldest working observatory in the world. It was built under the patronage of France's Sun King, Louis XIV, and inaugurated in his presence in a grand ceremony on June 21, 1667. The observatory has been the pride of the nation ever since, and the French have invited the world's greatest astronomers to work here. Among the most renowned was Giovanni Domenico Cassini (1625–1712), who came here from Italy and made important discoveries in the sky from the telescopes of the observatory. His name is known to anyone with familiarity with astronomy: The division between the rings of Saturn is called the Cassini division. Cassini first observed the gap in Saturn's ring in 1675. And the people of Paris remember him as well: The street that crosses the Avenue de l'Observatoire just in front of the observatory is rue Cassini. The main hall in the observatory, just below the dome, is the Salle Cassini (Cassini's Hall). This hall will figure prominently in our story. The Danish astronomer Ole Roemer (1644–1710) was also invited to work at the Observatoire de Paris. Here, in 1676, he measured the speed of light. Roemer's estimate of 140,000 miles per second was very good, considering the year it was made. The actual value is about 186,000 miles per second. Foucault would get close to the actual value of the speed of light in his own experiments in the second half of the nineteenth century. Foucault had been fascinated by Roemer's work two centuries before his time. And so was another person, who had risen to great prominence in French science—François Arago.

• • •

François Arago (1786–1853) was tall and handsome, sociable and easygoing, and had an enviable degree of self-confidence. He belonged to an illustrious family that was to give France some of its greatest leaders in politics, law, letters, and science. In fact, the French nation would rarely see a family contribute so much to its glory.

Arago was born in the small town of Estagel in Roussillon, in France's region of the Eastern Pyrenees, in 1786, three years before the French Revolution. Years later, while he was waiting to take an exam at the prestigious École Polytechnique in Paris, a fellow student destined to become one of France's most famous mathematicians, Adrien M. Legendre (1752–1833), asked him, because of his foreign-sounding name, whether he was "really French." In response, Arago told Legendre his family history. There had been Aragos in the French Pyrenees for centuries, and the family had always been active in public service. François's father, Bonaventure Arago, was the mayor of Estagel, and his parental ancestors for generations had held elected positions in the town council, regional government, and provincial civil service.

The name Arago is of Spanish origin and is related to Aragon. The people of this mountainous region are in their culture, habits, and dialect influenced by neighboring Spain; but they have always been French. With François and his siblings and children, France was about to receive a new generation of Aragos, one whose contribution to the nation would be considerable.

While he was always encouraged as a child to grow up to serve his people in politics, the young and clearly gifted François surprised his parents by announcing that he wanted to study mathematics.

France has one of the oldest university systems in the world. The famous Sorbonne—now one of the campuses of the University of Paris—was founded as a school of theology in 1253 by Robert de Sorbon under the illustrious king Saint-Louis (Louis IX). Later, philosophy and arts and sciences were added. But in 1530, King François I, "The Knight King," as he was called, decided to open a new institute of learning, the Collège de France, in which subjects that were not taught at the Sorbonne would be studied in a setting that allowed unfettered exploration of ideas and knowledge. He named six Royal Lecturers as professors in this new institution: two of Greek, three of Hebrew—and one of mathematics. Thus in 1530, mathematics became an important field of study in France. France's first "official" mathematician, the person the king appointed to the chair of mathematics at the Collège de France, was Oronce Finé (or Fine; it is not clear from the old records which kind of "e" is the last letter of the name). He was known for making clocks, and his specialty was geography and mapmaking. Finé, in fact, drew the first known map of France. His mathematics was an odd mixture of ancient Greek thought with medieval concepts, for he spent his time trying to square the circle (a Greek problem of antiquity that would be proven impossible within two centuries of Finé's time) using the mysterious Golden Section.

At the time of the French Revolution, two important schools of science and mathematics, as well as other fields, with no religious affiliation, were inaugurated. These were France's Grandes Écoles: the École Normale Supérieure and the École Polytechnique. In these institutions, France's most gifted mathematicians were trained and taught.

In the two centuries since the first professor of mathematics took his position at the Collège de France, mathematics had come a long way in France. There were Fermat and Pascal and Descartes, with their great achievements in mathematics and physics, as well as many others. The importance of mathematics as a tool for understanding the real world was widely accepted. Galileo's statement that "the book of nature is written in the language of mathematics" had become so meaningful that when Napoléon Bonaparte embarked for Egypt in 1799, he brought along with him two of France's best mathematicians, Joseph Fourier (1768–1830) and Gaspard Monge (1746–1818).

Monge had studied at a military school and solved a key problem of fortification by inventing descriptive geometry: a geometry in which three-dimensional objects can be described by their projections onto two perpendicular planes. Because of the importance of this work in military operations, Napoléon chose to take Monge on the expedition. Fourier had also done important work in applied mathematics. He laid the foundations for a theory of the conduction of heat and discovered methods that are now indispensable in modern physics, electrical engineering, and other fields. Napoléon recognized the importance of Fourier's mathematical methods and asked him to accompany him as well. In Egypt, the two French mathematicians had a unique opportunity to apply mathematics in the field. Monge, who was fifty-three, took part in designing fortifications and in actual fighting. Once, taking measurements on a boat on the Nile, Monge was almost caught and killed by the enemy when his boat hit a shoal. Napoléon saw him and galloped over with his troops to save the mathematician.

Leaving Egypt to return to France, Napoléon asked Fourier to remain behind to manage the Egyptian Institute, a scientific institute Napoléon had founded. Fourier used science to date the trove of archaeological finds the French had discovered in Egypt. When, some time later, Fourier's findings were made public in France, he incurred the ire of the Church because the age he assigned to some of the finds far exceeded estimates of the age of the Earth itself based on reading the Bible.

But along with the application of mathematics as a practical tool for analyzing problems of the real world, at the start of the nineteenth century, in France and elsewhere, a new kind of mathematics was emerging. This was the rebirth of "pure" mathematics, as the ancient Greek philosophers and mathematicians saw it: a discipline within which rigor in proof and details of logical implication were of paramount importance.

Ironically, within this new framework, the mathematicians of the nineteenth century would not have looked favorably upon the work of Newton over a century earlier had he been one of them. For Newton—while gifted beyond description—had not used carefully defined concepts nor followed rigorous rules of mathematical logic in his development of the calculus. Newton created (or discovered) an entire mathematical theory, one whose wide range of applications is too long to be listed and whose importance in today's world is immense. But he achieved this development using notions that were imprecise. One of them was "fluxion," a term that would make a mathematical purist shudder, since it was neither rigorously developed nor perfectly well defined. (Newton's "fluxion" would in time be replaced by the modern concept of a differential, leading to the derivative.) Yet Newton (and, independently of him, Leibniz in Germany) created the calculus, and the calculus does work wonderfully, regardless of how imprecisely or unrigorously—from the later perspective of nineteenth-century mathematics—it was originally developed. This new frame of mind within the schools of mathematics in France of the 1800s was dangerous. For if a talented modern Newton were to appear—someone who could derive wonderful new truths that could promote our understanding of nature but would do it without "mathematical rigor"—his or her achievements would be met with resistance. It was exactly this dangerous new attitude of mathematicians that in time would stand in the way of Foucault.

The new approach to mathematics was already evident in 1807, when Fourier, back from Egypt for several years, presented to the French Academy of Sciences his masterpiece, *The Mathematical Theory of Heat*. In this paper, Fourier proposed a theory to explain heat conduction using differential equations. This pioneering work is considered very important today. But Fourier was criticized by mathematicians, among them Laplace, Joseph Louis Lagrange (1736–1813), and Legendre, who argued that his work—while correct in its applications to physics—left out many important mathematical details, which they considered essential for a good mathematical theory. The reviewers claimed that Fourier's work lacked rigor, and therefore was defective—despite the fact that it was applicable in the real world, producing excellent predictions of natural phenomena and enhancing our understanding of the physics of heat. It took much work before Fourier could correct his work to satisfy the demands of the theoretical mathematicians.

Young François Arago was not encumbered by such problems. His talents, both in mathematics and in other disciplines he studied in school, were enviable. He was ready to begin his university education and was looking forward to the great new challenges and triumphs ahead. The world was his oyster.

The École Polytechnique was founded in 1794 by the Convention (one of a sequence of parliaments put in place by the French Revolution), following a suggestion by Gaspard Monge. Students were young, some as young as fourteen, and they lived outside the institution. The first selection of students included some who would become famous after their graduation: the physicist Jean-Baptiste Biot (1774–1862); Étienne Louis Malus (1775–1812), the man who in 1808 discovered the polarization of light by looking through a crystal at sunlight reflected from a window of the Luxembourg Palace; and the mathematician Louis Poinsot (1777–1859). They were followed a few years later by the chemist Joseph Louis Gay-Lussac (1778–1850) and the mathematician Siméon Denis Poisson (1781–1840).

Upon his return from his celebrated expedition to Egypt with Napoléon, Monge's first task as director of the school was to reorganize the curriculum to emphasize mathematics and physics. Shortly afterwards, his brother Louis Monge went to Toulouse to examine candidates for admission. Among them was seventeen-year-old François Arago. Arago tested very highly and, depending on sources, was either the best of all tested that year or at least one of the top six.

In 1803, he began his studies at the École Polytechnique, then housed at the Bourbon Palace in Paris. Some years later, Napoléon moved the school to the medieval buildings of the old Collège de Navarre, founded in 1315, on rue Descartes—just below the Panthéon.

Arago lived the exciting life of a young student in Paris, studying mathematics at the kitchen table of the apartment he shared with its owner, a friend of his father. That friend introduced him to the mathematician Siméon Denis Poisson, who was a few years older and already a professor of geometry. (Poisson was a mathematician, but often mathematicians liked to be called "geometers," for the reason described earlier.) Poisson had been a student of Pierre-Simon de Laplace, a defining figure in French mathematics who had addressed issues of the mechanics of rotating bodies and whose work Poisson extended. Poisson had been such a brilliant student that he was able to find errors in his professors' theorems. He was therefore given much leeway at the university and allowed to opt out of taking required courses he didn't like, such as graphical geometry. By 1807, when he was only twenty-five, Poisson was a full professor at the École

Polytechnique, and in 1811, he was already a member of the Academy of Sciences.

Arago and Poisson quickly became friends—they shared much in common, including rural origins, liberal political views, and a love of mathematics. Years later, Arago would recall that the young professor used to come to his room in the evenings, spending hours with him discussing mathematics and politics. Poisson, whom posterity would remember as one of France's most renowned mathematicians, thus became the young Arago's mentor and would lead him on a lifelong course of both science and politics.

But only two years after Arago began his studies at the university, an incredible opportunity came his way, one that would have called for a much more thoroughly trained individual, perhaps with a doctorate in science and years of experience—certainly not an undergraduate student. Nineteen-year-old François Arago was asked to measure the meridian of Paris.

In 1792, the Convention entrusted the two astronomers Pierre Méchain (1744–1804) and Jean-Baptiste Joseph Delambre (1749–1822) with the charge of measuring the meridian of Paris, a task that—once completed—would also have yielded the exact measure of the meter, defined to be a tiny fraction, one-forty-millionth, of the total circle from pole to pole and back around the world. The two astronomers went through a harrowing series of adventures on their way through France and Spain, one of them dying from malaria, and the task was still not complete.

Now Poisson, who had been assigned to find a successor to the unfortunate pair, offered the mission to his protégé. He introduced Arago to the leading French mathematician of the day, his own former professor, Pierre-Simon de Laplace. The latter was impressed with the young student. He enthusiastically supported Poisson's suggestion and encouraged Arago to accept the offer.

Arago was stunned, but kept his composure. After some thought, he agreed to assume this impossibly difficult, yet potentially immensely prestigious, mission under the condition that afterwards he be allowed the opportunity to satisfy all his other obligations, including finishing his studies and serving in the French artillery.

On February 22, 1805, Arago was formally nominated to measure the Paris Meridian. He moved to his new rooms in the annex to the Paris Observatory, on what today is the Avenue Denfert-Rochereau. Leaving his rooms and following the path on the leafy grounds of the Observatoire, he would soon find himself inside the majestic seventeenth-century edifice that France's last great king had built for science. A long flight of stairs would lead him to the high-vaulted Meridian Hall (also named Cassini's Hall, after the great astronomer). Meridian Hall is a long room, lengthwise perfectly aligned along a north-south axis; it has stark, bare yellow walls, arched windows, and a small round opening in the ceiling. The arched windows stretch upward, tall and narrow. They make the room seem like a Turkish bathhouse as seen through the eyes of El Greco. The Paris Meridian runs precisely through the center of this large room.

Meridian Hall (Salle Cassini). *(Courtesy of Josette Alexandre, Observatoire de Paris)*

Objectively, of course, there is nothing special about this location. Meridians are defined by people, as they attempt to identify locations on Earth for use in maps and navigation and to provide measurements of distance and time. A place has to be chosen arbitrarily through which to start drawing the pole-to-pole arcs we call meridians, or longitudes. Latitudes, also called parallels, are circles that go around the globe, from the equator to the poles, all parallel to the equator (and to one another). Until Greenwich took over because of Britain's dominance in worldwide navigation, the Paris Meridian was the "first meridian"—the one from which all other meridians, east and west, were measured.

In 1809, Arago completed the measurements that defined the Paris Meridian, extending it northward to the coast, as well as to the south, into the Balearic Islands and North Africa. But there is something magical about the way the city of Paris is built: for the meridian runs through the greatest treasures of the French capital. How could someone without a compass trace the Paris Meridian through the city? In the year 2000, a Dutch artist by the name of Jan Dibbets decided to commemorate Arago's achievement almost two centuries earlier in an unusual way. He made 135 brass medallions, about 6 inches in diameter, each bearing the name "Arago" and a marking for north and south, and embedded them in the ground all over Paris along the meridian, north and south of the Observatory. A delightful game for a visitor to Paris is to search for these medallions, seeing the best of Paris while doing it. (To give just a few hints: Try the Luxembourg Garden, the Louvre, the Palais Royal, sidewalk cafés, and even the quays along the Seine.)

So how did he do it? How did the nineteen-year-old Arago measure that imaginary line passing through the heart of Paris,

north to the coast and the North Sea, and south into the Mediterranean, the Balearic Islands, down through the African continent?

At the observatory, Arago came in contact with the best of
France's astronomers, physicists, and mathematicians. In a short
period of time, he learned how to make observations and how to
interpret them, as well as how to apply the methods of physics
and mathematics he learned from the famous mathematicians
Gaspard Monge, Pierre-Simon Laplace, Adrien Legendre, and
Joseph Lagrange. He also met Jean-Baptiste Biot, a young professor of physics at the Collège de France. Biot was chosen to
work with Arago to continue the measurement of the meridian.
Their measurements were required at the area where their predecessors had stopped: Spain. The project of measuring the
meridian had expanded in scope since its inception during the
French Revolution. French scientists now also wanted to know
the exact shape of the world. For this purpose, Biot and Arago
would use a pendulum. Finding the exact length of a pendulum
necessary for an oscillation of exactly one second at any given location would allow them to estimate the force of gravity at that
location. Scientists hoped that the collection of measurements obtained this way would give them an idea about the exact shape of
the Earth. Years later, Foucault would become familiar with
Arago's measurements using a pendulum, inspiring him to experiment with this device. On September 3, 1806, Biot and
Arago left for Spain.

The measurement of the meridian consisted of triangulation. The first few measurements of distance were made on the

ground, and all the others were carried out by measuring angles of triangles whose points were set to the tops of hills or other high landmarks, and the distances were computed using trigonometry.

Arago's memoirs are full of anecdotes about the adventures the pair of French scientists had in Spain, sharing their encampments with rats and brigands who demanded shelter and left in the morning without harming them; being caught in between a jealous lover and his fiancée; and Arago being punched in the mouth by the Archbishop of Valencia for an unintended, misconstrued impoliteness. But then, in 1808, while the pair was continuing their measurements on the island of Majorca, the real troubles began as France and Spain went to war. Here were two French citizens, carrying sophisticated measuring instruments, in the midst of Spanish territory. They were immediately assumed to be spies. Biot managed to return to Paris, but Arago remained behind to continue the work, disguised as a Spaniard with the help of local villagers. Eventually he was arrested as a spy and imprisoned together with a French officer named Berthemie. At home he was presumed dead, and the scientific community in Paris mourned him.

In the summer, the two Frenchmen managed to convince the commandant of the citadel in which they were being held that they were innocent, and he and his guards looked the other way, allowing the pair to escape. On June 29, 1808, they boarded a fishing boat headed for Algiers. The French consul in Algiers furnished Arago and Berthemie with forged Austrian passports, and ten days later they boarded a boat for Marseille. The boat also carried two lions as gifts from the local Arab leader, dey El Ghassal, to the Emperor Napoléon. As fate would have it, a Spanish corsair captured the boat, and the two Frenchmen again found themselves captives in Spain.

Throughout his captivity, Arago guarded his logbook containing his measurements with his life. Fortunately, none of his captors ever tried to separate him from his scientific treasure. At one point, however, he was forced to sell his watch in order to survive. One story has it that the watch ended up in his hometown in the Eastern Pyrenees and that his mother recognized it in the possession of a friend who had bought it from a traveling merchant, and she thus believed that this was the final proof of her son's demise.

As French forces advanced in Spain, the Spanish authorities transferred their French prisoners to Algiers. Dey El Ghassal, whose boat had been taken by pirates, and who was furious about the news that one of his lions had been killed in the attack, recognized the prisoners and immediately ordered that they be released and brought to his palace. He arranged for them to board another boat headed for Marseille, but a severe storm caused it to veer off course and end up back on the North African coast. The two Frenchmen were now imprisoned by hostile Moslems, who viewed them as infidels. Attempting to escape once more, Arago claimed to convert to Islam and was allowed to proceed toward Algiers. Unfortunately, dey El Ghassal had died and his successor was hostile to France. Since French shipping agents owed duties to the government of Algiers, the new dey took Arago and Berthemie, who had also ended up in

Algiers, hostage in lieu of these payments. The two French prisoners were about to be shipped off to a penal colony. But at the last minute, the French consul again intervened on their behalf, negotiating with local merchants to allow the two Frenchmen passage on a ship carrying cotton bound for Marseille and in return paying all the duties to the authorities.

On July 2, 1809, almost three years after he left Paris on his mission to measure the meridian, Arago arrived in France. He landed in Marseille and immediately booked passage by train to Paris, with all his observations of the meridian intact. He arrived in the capital to a hero's welcome. His career was launched. The Academy of Sciences was so impressed with Arago's incredible achievements that it immediately made him a member.[25]

Arago quickly rose in importance and influence in the Academy of Sciences and over the years would dominate the discussions and the agendas of this scientific body. He would do the same at the Bureau of Longitudes (a prestigious scientific body) and at the Paris Observatory. And on the political stage, he would become a key member of the French National Assembly and would serve for a time as Government Minister.

Upon his return, Arago became interested in the problem of measuring the speed of light. There is a connection here with his work on measuring the meridian, since the meter can also be defined—and is, in fact, today—as a set fraction of the distance light travels in one second. Arago began a series of experiments designed to measure the speed of light. He also wanted to know if light travels at different speeds through different media, such as air and water. But Arago had become such an important public figure in France that he never had the time to complete these involved experiments. Within a short time he was Director of Observations (the official title of the head of the Paris Observatory); he became the Perpetual Secretary of the French Academy of Sciences; he was elected to the Bureau of Longitudes; and, most important, he was elected to the National Assembly—France's legislative body. His son Emmanuel also served for years along with him in the National Assembly.

It is difficult to list all of François Arago's contributions to science, to the administration of scientific establishments, and to France and its institutions. The proceedings of the French Academy of Sciences include hundreds of notices, discussions, and presentations by Arago over many years. Equally, newspaper reports of France's National Assembly from the mid-1800s contain many references to bills introduced and discussed by François Arago and by his son.

Arago was getting old, and his eyesight was failing. He dreamed of having "young eyes" to complete his old experiments designed to measure the speed of light. Léon Foucault expressed great interest in working on this project. Arago had met the young Foucault in a number of his capacities. He also knew Hippolyte Fizeau.

When Daguerre first started looking for government support for his projects in 1835, he contacted Arago as an important member of the French National Assembly, in hopes of securing his help in gaining funding for his projects. Arago became fascinated by the idea of photography and used his influence to help Daguerre. In January 1839, Arago organized a demonstration of the magic of daguerreotypy to the French Chamber of Deputies. The demonstration helped convince the skeptics that this was a viable commercial project, and as a result, the French government bought the patent in August 1839 and put daguerreotypy in the public domain.

Later, as the project moved forward, Arago became aware of the further development of daguerreotypy by Foucault and Fizeau and made the acquaintance of the two young medical students interested in science. When Donné's and Foucault's *Atlas of Microscopy,* which featured the daguerreotypes of objects seen through a microscope, was published, Arago was impressed by the high quality of the pioneering use of photography in science and wondered whether similarly successful results might also be achieved using daguerreotypy through a telescope. Coincidentally, Foucault was moving in the same direction. He became interested in astronomical instruments and began to experiment with a heliostat, an instrument for viewing the Sun. All that remained was to put the two interests together: to use the heliostat to produce a daguerreotype of the Sun. And this indeed was one of the projects that François Arago had in mind. In 1845, Arago contacted Foucault and Fizeau, and the pair took the first ever photograph of the Sun. The image they obtained was of exceptionally high quality, and it clearly shows a number of sunspots.

Arago was very pleased with the results and encouraged

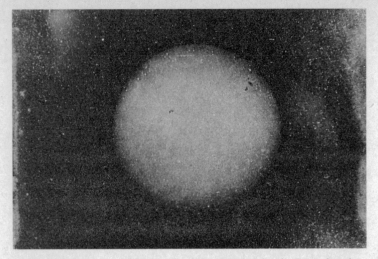

The first photograph of the Sun, the daguerreotype taken by Foucault and
Fizeau in 1845. *(CNAM, Paris)*

Foucault and Fizeau to continue their work. He thought that
perhaps these two young researchers would be the right people to
continue his important project of measuring the speed of light.
That was the same year that Foucault took up the role of reporter
for the *Journal des Débats.* Arago, who was the Perpetual Secretary of the Academy of Sciences, thus saw Foucault every week at
the meetings of the Academy of Sciences. He appreciated the
young man's reports to the public on a wide variety of scientific issues and recognized his boldness and ambition. He became convinced that Foucault and his friend were indeed the right people
to assume the complicated task of measuring the speed of light.

Foucault and Fizeau brought much energy and enthusiasm to
their new scientific task. Both knew that measuring the speed of

light would not be easy and that they would first have to learn much about the technology available to them, as well as about the many previous attempts over the centuries to measure this elusive constant.

Even before Roemer's experiments at the Paris Observatory in 1676, Galileo attempted to measure the speed of light early in the 1600s using lanterns placed on two hills in Tuscany. When his assistant uncovered a lantern on one hill and Galileo saw the light, he in turn uncovered his own lantern, and the assistant noted when he saw the light. Of course the distance between two hills was too small and the speed of light so great that this quaint experiment did not yield any results. Roemer, however, had obtained good results for his time (230,000 kilometers per second, as compared with the true value of roughly 300,000 kilometers per second), and Arago had made progress. The nineteenth century, with its tremendous advances in engineering and technology, offered an opportunity that could not be missed to measure the speed of light. For the first time, it was possible to spin a wheel at precisely known high speeds in the hundreds or even thousands of revolutions per second—which was exactly the technology required for this measurement.

The motivation for the study of the speed of light was a desire to understand the nature of light. Today we know from quantum theory that light can be viewed as both wave and particle, but before the emergence of quantum theory, scientists kept attempting to determine which of the two theories was correct: the particle theory or the wave theory.[26] Arago believed that measurements of the speed of light in air and in water might settle the problem definitively. He came to this conclusion because he knew that according to the particle theory, light would speed up as it entered water from the air, while according to the wave theory, light would slow down once it entered the water. He thus designed an experiment that would test the relative speeds of light in the two media.

Arago's design used a single beam of light split in half. One ray went through the air, and the other traversed through water. A mirror spun at 2,000 revolutions per second then reflected both beams of light. The slower one would be reflected at a larger angle. But Arago's deteriorating eyesight had prevented him from determining the result, and he discontinued the experiment.

Now Foucault and Fizeau resumed Arago's interrupted project at the Observatory. First, they worked together, reassembling Arago's instruments and experimental setup. But suddenly the two friends had a falling out. For four years they had studied science together and achieved much. According to Foucault's chronicler Lissajous, despite the fact that Foucault and Fizeau had built a common fund of knowledge through their joint work, each man had his own ideas and wanted to pursue them alone.[27] From collaborators, the two young men became competitors. Fizeau continued on Arago's track, using the old scientist's apparatus and technique. He also pursued his own research on determining the absolute speed of light. In a now-famous experiment in July 1849, Fizeau used two lenses separated 8,633 meters from one another: One was located by his parents' house at Suresnes and the other on Montmartre, the high hill on the Right Bank in Paris, which during this time was not settled and was covered with vines. Between the lenses were mirrors and toothed wheels rotating very fast. His result was 315,000 kilometers per second, which was closer to the actual value than any estimate obtained until that time. This estimate was 5.1 percent higher than the value of the speed of light we know today. As we will see, the next estimate of the speed of light would be obtained a few years later by Foucault, and his error would be ten times smaller.

Foucault continued to work exclusively on the project of determining the relative speed of light in air and in water, and here chose his own route. He was eager to achieve a good result, now that his friend and new competitor had won recognition for his experiment to estimate the absolute speed of light.

Foucault built a small steam engine and used it to drive a mirror at a lower speed of 800 revolutions per second. His experimental setup was only 4 meters long and consisted of a spinning mirror and a stationary one. Sunlight from his heliostat was reflected first by the spinning mirror and then by the stationary one. Because light travels at a limited—albeit very high—speed, and one mirror spins fast, the reflected light does not arrive at its starting point, but rather is deflected somewhat. This relative deflection can be measured when air separates the two mirrors and also when a transparent tube of water is inserted between them. In April 1850, Foucault successfully completed his experiment, proving that light traveled slower in water than in air, as predicted by the wave theory of light.[28]

In describing his work, in characteristic modesty, Foucault wrote:

> We have invented neither spinning mirror, nor achromatic lens, nor network, nor micrometer; we have had the good fortune to group these instruments acquired by science in a manner that furnished us with the solution of a problem posed twelve years ago.[29]

The composer Hector Berlioz (1803–1869) was also a journalist for the *Journal des Débats*. While Foucault was carrying out his experiments on the speed of light, Berlioz wrote him a letter asking him if he could bring some friends over to observe Foucault's experiments.

Seven weeks after Foucault completed his experiment, Fizeau obtained similar results on the relative speed of light in air and in water using Arago's original design, confirming his friend's result. François Arago was elated by the results of Fizeau and Foucault. He especially valued Foucault's discovery that light travels slower in water than it does in air, since it led to what he considered a very important conclusion about the nature of light. According to Joseph Bertrand, Arago's reaction was the following.

> Arago smiled at the beautiful experiment, which pleasantly evoked, by its well-deserved praise, his own days of glory when, beating Laplace, Poisson, and Biot, he entered the Academy of Sciences.[30]

The Perpetual Secretary made the Academy of Sciences aware of the discovery. But Foucault was not invited to join this scholarly group. Foucault knew well that, if nothing else, the results of his work on the speed of light should have been enough to earn him a Ph.D. in physics. Arago, however, was impressed with him, and this fact was very important to the aspiring young scientist.

So during that long night in January 1851, Foucault thought of Arago. The old Director of Observations at the Paris Observatory and Perpetual Secretary of the Academy of Sciences, not to mention senior member of the National Assembly, was the one person in the world who both appreciated his talents the most and was in a position to enable him to bring his newest discovery to light. Arago would help him, he thought as he finally fell asleep.

6

---◆---

"COME SEE THE EARTH TURN"

As Foucault later described it, Arago kindly agreed to allow him to present his pendulum at the Observatory. Thus the largest, highest, and most famous room in the Observatory, Meridian Hall (Salle Cassini), was put at Foucault's disposal.

Foucault immediately set to work. He transported the sensitive mechanism he had devised, which allowed a pendulum to swing in any direction without torque on the line, from his basement on the rue d'Assas down to the second point of our imaginary triangle mentioned in the Preface: the Observatoire de Paris. Foucault climbed up to the opening in the ceiling of Meridian Hall and carefully installed his mechanism. Meridian Hall was much higher than his basement—it allowed for the use of a pendulum 11 meters long.

Foucault had much riding on this experiment, and he

wanted it to proceed perfectly—no snapping wires—and with great precision. Accuracy would be of paramount importance for proving that the pendulum reflected the motion of the Earth underneath it. So, at his own expense, he hired the best crafts-man he could find: Paul Gustave Froment (1815–1865), a man who was well known in France for his high-quality work with brass and other metals. It was important to have a perfect pen-dulum, hung just right, and started on its motion very carefully, so that its natural movement would not be disturbed by the human hand. For otherwise, the motion of the plane of the swing of the pendulum could be blamed on the initial condi-tions of motion. Froment produced such perfect pendulum bobs that people still marvel at how they look and perform today. His pendulums made for Foucault are now displayed at the museum of the Conservatoire National des Arts et Métiers (CNAM) on rue Saint-Martin in Paris.[31] In the actual experi-ment, Froment would burn a woolen thread that secured the pendulum to the wall, so that it would start to swing with no perturbation from human touch.

Foucault and Froment checked that the apparatus was in order. They performed a few trial runs in the high-vaulted hall with its arched windows, and the pendulum—its center aligned with the Paris Meridian passing right underneath it—was ready to go.

It was now time to write the invitations to this great scien-tific demonstration, one that would—Foucault hoped—establish him as a scientist of repute. For, after all, Foucault had by now invented a revolutionary lighting technology used in science and the theater; he had measured the speed of light and proved it was lower in water than in air; and he had come up with the first piece of irrefutable evidence that the world turns. He deserved credit for all of these contributions, and this was his great opportunity to impress the savants of the Académie des Sciences, the mathematicians and scientists who carried the torch of French scholarship and research.

He made invitation cards, and wrote on each one:

You are invited to come to see the Earth turn, tomorrow, from three to five, at Meridian Hall of the Paris Observatory.[32]

On February 2, 1851, Foucault sent this invitation to all the known scientists in Paris.

It is hard to know what Foucault's patron, Arago, thought. Quite possibly, once he gave Foucault the permission to go ahead and use Meridian Hall for his demonstration, he had little time to pay attention to what Foucault was doing. François Arago was the busiest man in France. He held three full-time jobs, each of which would have exhausted an ordinary person. But the man who had measured the Paris Meridian was no ordinary person. And he performed his multitude of tasks perfectly well even though he was suffering from poor health and deteriorating eyesight.

Arago spent his days and nights at three of the most impressive locations in all Paris. One was the Observatory, of which he was the director and at which he had to supervise and organize and administer many research projects at the leading edge of astronomical science.

His second job, as a representative of the French people, led him to spend hours and days at the National Assembly building, which still sits on the bank of the Seine across the Concorde bridge (made with stones from the old Bastille) from the obelisk of the Place de la Concorde, where the guillotine once operated during the Revolution. The National Assembly was, and still is, an impressive building with high columns and crowned by the statuary by Cortot of France flanked by figures representing power and justice. Legislative meetings at the Assembly were long and arduous, and Arago and his son and the other representatives had recently had an added burden: that of dealing with a new President of the Republic, a man who acted like a prince and, in fact, was one. This prince was to play a key role in our story.

Arago's third job was that of Perpetual Secretary of the Academy of Sciences. This was the title of the director of the august body that saw to the advancement of the sciences in France. The Academy has an illustrious history.

The French Academy of Sciences was founded in the seventeenth century. On December 22, 1666, seven mathematicians and seven physicists met at the king's library to inaugurate the Academy. The inspiration for scientists to form such a society was the informal fraternity of scientists who had gathered around Marin Mersenne, the monk and friend of Descartes, who lived in the Minim monastery at the Place Royale (today Place des Vosges) in Paris and who had a love for science and mathematics. Mersenne, in fact, created the idea of a community of scientists, and modern science owes him much, because today science is often pursued by groups rather than by individuals. Through his 10,000 letters, Mersenne had connected Fermat, Descartes, Huygens, Torricelli, and others, allowing them to coordinate research efforts and to help and learn from one another.

The fourteen scientists who met in 1666 in the royal library sought to continue this tradition and work together to further knowledge. The Sun King, Louis XIV, who had also inaugurated the Paris Observatory, gave the new Academy of Sciences his formal decree of protection in 1699, endowed the body with a constitution, and retained for himself the right to appoint members. The membership later became chosen by election, and it included France's greatest mathematicians and scientists. In 1805, the Academy, which until that time held its meetings in the Louvre, moved to its new home across the Seine, in the building of the Institut de France. The Institute was inaugurated as a federation of five academies, including the academy responsible for the French language. But, over time, the Institute came to be dominated by the Academy of Sciences.

The stately building of the Institut de France and the Academy of Sciences was built in 1662 to house the library of Cardinal Mazarin. Its gilded dome still rises over the river, and below it, arranged in a semicircle facing the Seine and the Louvre, are the hallowed halls of the Institute—the libraries, archives, and meeting rooms for the exclusive use of the members of the Institute and its Academies.

In 1835, François Arago inaugurated the *Proceedings of the Academy of Sciences*. This important publication helped disseminate the scientific findings and discussions of the Academy throughout France and the world, enhancing the prestige of this body of scholars. In 1851, for example, the year of Foucault's pendulum experiments, the *Proceedings of the Academy of Sciences* contained a wealth of new knowledge and scientific discussions. These included an article on the new discovery of eggs of an unknown giant bird in Madagascar; a report by Fizeau on research on the propagation of electricity carried out in the United States in 1848 and 1849; a report by François Arago on the interference of polarized light; a short communication about a study of the results of the new vaccine for yellow fever; an article by Charles Bonaparte, a scientist relative of Napoléon, on a newly discovered bird on the banks of the White Nile; and a correspondence from the French Minister of War about his report to the President of the Republic on issues relating to the administration of Arab tribes in Algeria. But the majority of articles in the *Proceedings* were lengthy, abstract, and rigorous mathematical derivations by the mathematicians Liouville, Hermite, and Cauchy (all of them names familiar to anyone who has studied mathematics). Baron Augustin Cauchy (1789–1857) was the most prolific of them all. Over the course of this year, he published in the *Proceedings* an entire theory of functions of complex variables, developed with great mathematical detail.

But the honor, respect, and renown heaped on members of

the French Academy of Sciences had a predictable downside as well. Members became enchanted with their own greatness; they were, after all, at the top of French science and of French society as a whole. They began to believe in their own infallibility and tended to dismiss the work of those who were not members.

An example of this frame of mind is given by the shameful story of the young mathematical genius Évariste Galois (1811–1832), who before the age of twenty had discovered an entirely new and beautiful theory in mathematics, known today as Galois theory. Galois sent his theory to Augustin Cauchy at the Academy of Sciences. Cauchy lost the paper. In January 1831, Galois sent another copy of his paper to the secretary of the Academy of Sciences, asking that it be read by the members. Galois had hoped that the mathematicians of the Academy would recognize that he was proposing an entirely new theory in the field of algebra. Siméon Denis Poisson read the paper and prepared a report to the membership. His report said that he and his colleague Lacroix could not understand the paper and thus declined to give it their approval. A short time later, the desperate Galois died in a duel. Today we know that Galois's work was both correct and of great importance in mathematics—but members of the Academy could not even understand it.[33] The Academy's shameful treatment of an outsider was about to repeat itself twenty years later.

Both Cauchy and Poisson are well-known names in mathematics today. And both mathematicians are famous for highly theoretical mathematics—which, perhaps, explains their lack of interest in the work of people they did not *perceive* to be on their level. Cauchy is remembered by mathematicians for his work in analysis, and for the "Cauchy sequence" named after him. There are also Cauchy's theorem, the Cauchy-Schwarz inequality, the Cauchy conditions, and Cauchy's formula. Modern mathematical analysis owes much to his work.

Poisson, on the other hand, is well remembered in the realm of applied mathematics—however, for highly theoretical work as well. The Poisson process and Poisson distribution are named after him, and he is remembered for important results in the theory of random variables. As opponents to Foucault (through their writings), these two mathematicians would be formidable foes.

7

MATHEMATICAL BEDLAM

France's scientists and savants did come to the Observatory on February 3, 1851, and they did "see the world turn." Foucault's pendulum performed exceptionally well. There is an elegance to a large, heavy pendulum swinging slowly back and forth. This pendulum had the added advantage of not only swinging in a stately manner across the stark surroundings of Meridian Hall—it slowly shifted its orientation, rotating ever so slowly over the Paris Meridian. And the scientists gathered around this pendulum immediately understood what they were seeing.

Since its inception, the French Academy of Sciences has been dominated by mathematicians and physicists. Among the mathematicians and physicists watching the pendulum swing that afternoon, there was not one who doubted that they were indeed watching the Earth turn beneath the pendulum. They all enjoyed the demonstration, many referring to it later as Foucault's "belle expérience" (beautiful experiment). They were clearly impressed.

But soon enough, the questions arose: How was it possible that no one had thought of this before? The experiment seemed so incredibly simple. Why hadn't any of the scientists and mathematicians who had spent lifetimes studying rotations and gravity and astronomy thought of this experiment? The mathematicians were angry that their equations had not predicted this phenomenon, and the physicists were equally upset that their physical intuition and analysis never led them to the "beautiful experiment" demonstrating so clearly that the Earth rotates. More important, the question that begged an answer was: What does mathematics say about this experiment? Shouldn't the equations of motion, developed by generations of mathematicians and physicists (who were often the same people) from Galileo to Kepler to Newton to their own members, have predicted this phenomenon? Some of the members were already beginning to say: "But I could have told you so. It's all in the equations." And, in fact, many of them had over the years developed equations that dealt with rotational motion and moving bodies and the Earth. However, not one of them had thought up such an experiment; not one of them had predicted that a pendulum would exhibit such change in its plane of oscillation in response to the turning of the Earth. *Au contraire:* Some of them had claimed that this would *not* be possible. And not one of the mathematicians or physicists of the Academy had an equation or formula that would tell them how fast the pendulum's plane of swing must change at any given location on Earth.

Cauchy never thought it possible that a pendulum should change its plane of oscillation, and Poisson, as early as 1837, had said that a pendulum would *not* move in such a way. But there were now equations galore to explain the movement Foucault had just shown them.

How did Foucault know to look at a pendulum to find proof of the rotation of the Earth? Was it by chance? Foucault created his original experiment in his cellar knowing exactly what to look for and why. The wonderful thing about Foucault was that the man always knew what he was doing. He *planned* his experiment—chance had little to do with it. Foucault was a creator, a born engineer with golden hands. By the time he had turned his attention to the problem of the rotation of the Earth, Foucault had constructed, with his own hands, a whole array of technical implements. To measure the speed of light, he had built a device with a mirror that rotated 800 times a second. It was powered by a little steam engine he had built. Later he used a compressed air machine fueled by a bellows to drive the same rotating mirror. All of these ingenious wheels and engines taught him a great deal about rotating bodies. In particular, Foucault had noticed a curious phenomenon. He was working with a little steel rod that was attached to the chuck of a lathe. When Foucault observed the vibrating rod, he noticed that its plane of vibration did not rotate with the chuck. The reason for this phenomenon is that the inertial body, in this case, the rod, maintains its inertial motion, its vibration, as no force is acting on it. This is in accordance with one of Newton's laws of motion. Foucault knew that a pendulum would behave similarly. While the point of suspension of the pendulum rotated with the Earth, no force acted on the pendulum itself, and thus its plane of oscillation could not change as the Earth rotated under it. This combination of Foucault's engineering skills and his curiosity about nature brought him to his masterpiece—a proof of the rotation of the Earth. The idea was born of both developments in science and engineering. Foucault made no secret of the inspiration for his experiment. He reported it to the Academy along with his findings.

But Foucault himself, the "nonmathematician," as the members of the Academy thought of him, not only provided the proof of the rotation of the Earth: He had even derived an equation to describe it.

Already on February 3, the same day Foucault demonstrated his pendulum in the Observatory, François Arago read a communication by Foucault to the members of the Academy of Sciences. In this paper, Foucault discussed his first pendulum experiment, carried out the previous month in the cellar of his house, and the resulting proof of the rotation of the Earth. More importantly, Foucault presented his formula—now called the *sine law*—for determining the length of time it takes, *at any given latitude,* for the pendulum's plane of oscillation to sweep a full circle and return to its starting point.

At the north (or south) pole, it takes twenty-four hours to complete the cycle; on the equator, the plane of swing of the pendulum does not move at all. And at intermediate locations, the period is equal to twenty-four hours divided by the sine of the latitude.

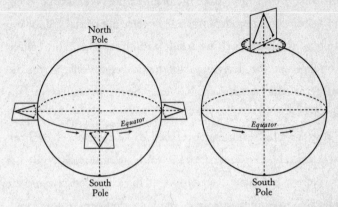

A pendulum at the equator does not change its plane of oscillation in response to the rotation of the Earth.

The plane of oscillation of a pendulum placed above the North Pole rotates around a complete circle once in 24 hours.

FOUCAULT'S SINE LAW

$$T = 24/\sin(\theta)$$

Where T=time required to complete a circle; sin is the trigonometric sine function; and θ is the latitude.

Hence in Paris, latitude 48 degrees, 51 minutes, north, it takes just under thirty-two hours for the plane of oscillation of the pendulum to return to its starting point.[34] (Note that the latitude of the pole is 90 degrees; sin (90) = 1, and hence T = 24 hours. At the equator the latitude is zero; sin (0) = 0, and hence T = 24/0 = "infinity," meaning that the pendulum's plane of oscillation does not change at all.)

This was an incredible finding, since it is not obvious why the sine function is the correct one to use in an expression to describe the time it should take the plane of the pendulum to complete a circle; and proofs of the sine law are not trivial.[35] Foucault had obtained this surprising result without mathematical training or experience in deriving mathematical equations, and he did it before the mathematicians had even begun to understand the problem.

But the mathematicians refused to be impressed by Foucault's formula. A week after Foucault's demonstration at the Observatory and the presentation of his paper describing the pendulum experiment and the sine law, members of the Academy scrambled to explain his experiment their way, as well as to protect themselves from criticism. They had all been put to shame by Foucault's achievement. Was it really possible that Foucault should discover something that the mathematicians' own equations did not predict? And how in the world would someone with no training in mathematics develop a law describing how fast the pendulum's plane must rotate for a pendulum placed *anywhere* on the planet? Foucault only had one observation point: Paris. This was a stunning achievement for Foucault, they knew, and an equally embarrassing situation for them, the "experts."

The mathematicians were quick to try to respond to the shock of Foucault's incredibly simple demonstration of the rotation of the Earth. They made the problem very complicated by referring to their old equations, searching to find in them something that "should have told them" what Foucault's pendulum, majestically swinging and turning as the Earth rotated, was showing them so clearly and so simply.

The first to make a written presentation read in front of the Academy at its February 10 meeting was Jacques Binet (1786–1856). Binet started by saying: "Foucault has asked me in what way the mechanical result he had obtained accords with mathematical theories." He proceeded to quote a succession of results by mathematicians over generations:

> In Chapter 5 of Volume 4 of his *Celestial Mechanics,* Laplace deals with falling bodies, but not pendulums. Poisson dealt with this subject in 1837 in his article in the *Journal de l'École Polytechnique* (vol. 26). There he said: "This force acting on the plane of oscillation of the pendulum is too small to move the pendulum perceptibly, to have an appreciable influence on its movement." This fact is contrary to the experiments of Foucault.

Binet then continued to try to defend this embarrassing failing of his revered late colleague (Poisson had died eleven years earlier). It seemed he would stop at nothing, for this is what he said: "But the passage I just cited permits a doubt: Poisson doesn't report a calculation of the force of which he speaks, and thus it is insufficient to allow us to know whether the perturbing force is *very small* [his italics], to conclude that it will only produce an imperceptible effect after a large number of oscillations."[36] He then addressed the sine law—that stunning piece of mathematical deduction by the untrained Foucault. To try to understand Foucault's new law, Binet had to quote theorems and derivations by some of the world's greatest mathematicians. He started with a theorem by the legendary Swiss mathematician of the previous century, Leonhard Euler (1707–1783); he then continued on to the work by the French mathematician Joseph Louis Lagrange; and then went on to the work of Louis Poinsot. After stating Foucault's law as "The angular momentum is thus 15" x sin(γ) for a second of sidereal time (being 15 degrees in a sidereal hour)," he went through the mathematical derivations he would need to explain it: "There is a theorem by Euler, which Lagrange has developed in his *Mécanique,* and on which the theory of couples of Poinsot sheds light. . . ." But Binet stopped short of actually *proving* Foucault's result using all this lofty mathematics. For this, he would need even more time.[37]

The *Comptes Rendus* (Proceedings) for the February 10 meeting continue as follows: "On the occasion of Binet's note, M. Liouville [this is Joseph Liouville, 1809–1882; professor of mathematics at the Collège de France] exposed, *vive voce,* in detail, a synthetic method that seemed to him also *rigorous.* . . ." Thus the idea of mathematics having to be done rigorously (meaning with carefully detailed theoretical derivations always in perfect accord with the rules of logic) entered this discussion. The "pure" mathematicians were preparing to outdo the unlearned Foucault. Ironically, the *Comptes Rendus* give no indication of just what kind of rigorous treatment Liouville gave to Foucault's result. Based on later details, Gapaillard presents Liouville's proof of Foucault's sine law.[38] This proof is given in the

Appendix. It begins by considering two extreme cases: the North Pole, and the equator.

To be sure, Liouville's attempt at what he called a *rigorous* proof of Foucault's sine law, by looking at the extreme cases of the poles and the equator and then addressing what happens at intermediate locations such as Paris, was not even original. Foucault himself had used this approach in his original presentation to the Academy on February 3. Foucault's paper included the following:

> The very numerous and important observations of which the pendulum has until now been subject to have mostly been related to the duration of oscillations. The ones I here propose to bring to the attention of the Academy have to do principally with the direction of the plane of oscillation, which, changing gradually from east to west, furnishes a perceptible sign of the diurnal movement of the terrestrial globe. Before attempting to justify this interpretation, I would like to abstract the movement of the Earth, and to suppose that the observer transports himself to the pole to establish a pendulum reduced to its greatest possible simplicity.[39]

Foucault then went on to argue why at the pole the time required to complete a revolution was twenty-four hours. Foucault himself mentioned Poisson's 1837 paper, for its implications about falling objects—diplomatically avoiding the part about a pendulum not showing any perceptible effect of the rotation of the Earth. He ended his paper modestly pointing out that his pendulum experiment allows us to proceed from the domain of theory to that of experimentation. This was a nod at the theoreticians, even though Foucault didn't have to do it, since he himself had derived the theoretical sine law without the aid of the mathematicians' theories.

The Academy (on February 10) then heard from a certain M. Fermont, who told the membership that *he* had actually obtained similar results to Foucault "some years ago." He did not elaborate.

A week later, on its February 17 meeting, the Academy had in front of it a full report on Foucault's experiment by Jacques Binet. Binet's presentation consisted of nine dense pages, replete with differential equations, trigonometric functions, and various formulas, all of them required for a proof of Foucault's sine law. The report concluded: ". . . conforming with the theory of M. Foucault, which the experiment has confirmed."

Here were France's greatest scientists and mathematicians—giants who in previous generations were appointed by the king himself, members of an elite group of achievers—and in front of them was the work of a lab assistant who in the past had been allowed into their midst only so he could report on their great discussions and presentations in a newspaper read by the average citizen. And he, the untrained lab assistant with no mastery of mathematics, proved the impossible and derived the law it was now taking them so much effort, and requiring references to so many previous works by great mathematicians, to prove. Just how Foucault managed to do it was a great mystery. And the embarrassment to the members of the Academy was considerable.

The defense by mathematicians of their own against the work of an uneducated scientist went beyond the borders of France. On March 16, 1851, Jean Plana published an article in the *Proceedings of the Academy of Science of Turin,* in which he wrote:

> To explain the fact of the movement of the plane of oscilla-
> tion of the pendulum one must recourse to the differential
> equations that determine the relative movement of a mate-
> rial point situated close to the surface of the Earth. These
> equations have been given by Poisson in 1837, on p. 21 of
> his work *On the movement of projectiles.* There, he reached a
> point declaring such movement to be impossible, and said:
> "The perpendicular force on the plane of oscillation is too
> small to make the pendulum veer from its plane in a per-
> ceptible way, and as a result, this force will not have an ap-
> preciable influence on its movement." Clearly there is a
> contradiction here with the experiment of Foucault, but it
> is wrong to think it resides effectively in the theory of Pois-
> son and that the new phenomenon must make us modify
> that theory.

Plana then proceeded to analyze the mathematics of Poisson and concluded that at one place in his paper, Poisson had actually meant *cosine* when he wrote *sine,* and whence his false conclu-sion.[40] It seemed that the mathematicians would stop at nothing.

Joseph Bertrand summed up the circumstances of the proof of the rotation of the Earth as follows:

Let us say very clearly, for that is true, that the geometers had signaled the direction; and add, for that is just, that they had not explored it; that deplorably quickly, Poisson had judged it unworthy of attention; and that it was Foucault, without any help and without a guide, who was the first to advance it.[41]

This was Foucault's highest point with respect to the Academy. With one technical exception, Foucault's name was not mentioned again in the *Proceedings of the Academy of Sciences* in 1851. And when a report was published in the *Proceedings* later that year about a Foucault pendulum experiment in Brazil— which was very important since it verified Foucault's sine law in the southern hemisphere—there was no mention of the name Foucault. It was as if the man who invented the pendulum experiment and derived its law never existed.

After the initial flurry of activity following the experiment at Meridian Hall had died down, the man and his achievements seemed again destined for obscurity. Unbeknownst to him, however, help would come from an unexpected direction.

8

◆

A NEW BONAPARTE

Napoléon Bonaparte, once he became emperor and had most of
Europe under his control, worried about his succession. His
achievements would not be lost, he hoped, if he could establish a
dynasty—not unlike the Bourbons, which the Revolution that
put Napoléon in charge had eliminated. For that aim, Napoléon
made his brother, Louis Bonaparte, the King of Holland. He also
encouraged him to marry his mistress Josephine's beautiful
daughter, Hortense de Beauharnais. This marriage, which was
not a loving one by all accounts, made Hortense the Queen of
Holland.

On April 20, 1808, the cannon at the Invalides square in
Paris was fired to salute a royal birth: Queen Hortense had given
birth in Paris to her third child, a son, who was named Louis-
Napoléon. Two years later, in the Palace of Fontainebleau south

of Paris, Emperor Napoléon himself presided over the baptism of his new nephew.

After Napoléon's defeat at Waterloo, circumstances changed for the Bonapartes. The victorious powers forbade any of Napoléon's heirs to reside in France, fearing that one of them might try to resurrect the lost empire. As a result, Hortense, no longer the Queen of Holland, moved with her children to Arenenberg in Switzerland. There, her son, Prince Louis-Napoléon, grew up speaking his native French with a foreign accent. But the family never forgot its legacy, and Hortense raised her children to love France and to strive to return to their homeland.

Napoléon's own son, his direct heir who would, if he ever took power, be called Napoléon II, died in Schoenbrunn Palace near Vienna at a young age. Thus, Louis-Napoléon and his siblings remained next in line in the Bonaparte dynasty. This fact never escaped the mind of young Louis-Napoléon Bonaparte. As he was growing up, one memory always remained in his mind and fueled his ambition and drive to succeed and resume his family's lost glory: the vision of his uncle leaving for his last battle at Waterloo.

Once Louis-Napoléon's older brother died in battle in 1831, Louis-Napoléon was entrusted with keeping the family name and legend alive. France was governed at the time by Louis Philippe d'Orléans, King of the French, his branch of the French royals having taken over from the Bourbons and beginning a reign called "The July Monarchy." Louis Philippe jealously guarded his realm.

Monarchies were a precarious business in post-Revolution France, and Louis Philippe was concerned about possible threats to his rule, not only from the Republicans, but also from France's other dynasties: the Bourbons and the Bonapartes.

As the person carrying the torch of the Bonapartes, Louis-Napoléon was eager to challenge the King. The young Bonaparte gathered supporters, and on October 30, 1836, the twenty-eight-year-old prince launched an ill-conceived, haphazardly prepared attack on a military base in Strasbourg from across the border. The band of rebels gathered by Louis-Napoléon has been described as "a tiny group of exalted idiots."[42] Louis-Napoléon had hoped to rally the French troops at the base to join him in an advance on Paris. The attack failed miserably. Colonel Taillandier, the commander of the base Louis-Napoléon tried to take, closed the gates in front of the small band of attackers, cutting them off from the men inside, whom Louis-Napoléon was hoping to rally to his cause. He then galloped up to Louis-Napoléon and, in front of his troops behind the fence of the base, ripped Louis-Napoléon's uniform with his sword and placed him and his followers under arrest.

In an ironic twist, the message Taillandier dispatched to Paris was interrupted, so that only the first part, saying that Louis-Napoléon had attacked his base, arrived at the capital. King Louis Philippe spent the rest of the day fearing that Louis-Napoléon had been successful and was perhaps already on his way to Paris. Only that evening did the rest of the message arrive, saying that the prince had been arrested.

Louis-Napoléon did arrive in Paris, under lock and barrel,

following the failure of this juvenile adventure. King Louis Philippe didn't quite know what to do with him. If he imprisoned the young adventurer, he would risk giving a pulpit to the rival Bonapartists, and who could tell where that might lead. So instead, the king quietly put Louis-Napoléon on board the naval ship *Andromède* headed to an undisclosed destination. The ship sailed that evening heading west. When the captain opened his sealed sailing orders once the ship had traveled a significant distance into the Atlantic, he found that he was to sail to Rio de Janeiro, wait there for a month, then head for Norfolk, Virginia, and deposit the rebel on American soil. Louis-Napoléon was told to stay forever in the United States and never set foot in France again.

Once in the United States, Louis-Napoléon made his way to New York. There, he was befriended by Cornelius and Margaret Roosevelt (the grandparents of the future president, Teddy Roosevelt) and through them met the authors James Fenimore Cooper and Washington Irving. He adopted a leisurely lifestyle, occupying himself mostly with the pursuit of women. But a letter from his mother put an end to that. Hortense was ill, possibly with cancer, and was not expected to live much longer. She wanted to see her son one more time. On June 12, 1837, Louis-Napoléon boarded the *George Washington,* heading for England.

The French got word of the prince's arrival in Britain, and Louis Philippe immediately instructed his ambassador to send agents to trail Louis-Napoléon to make sure he did not attempt to land in France or anywhere on the Continent. Louis-Napoléon evaded these agents by checking into a hotel in Rich-

mond, a resort on the Thames, leaving it under cover of dark-
ness, and hurrying to the London docks, whence he boarded a
steamship for Rotterdam. From there, the prince made his way
to Arenenberg, to his mother's deathbed.

Louis-Napoléon inherited a significant amount of money and
property from his mother. He moved from Arenenberg to the
neighboring town of Gottlieben, not far from the French border,
and settled there. Soon, however, Louis Philippe began to feel in-
secure about a rebel living so close to his kingdom. The French issued a formal demand to the Swiss government to expel the
prince from their country. The Swiss refused, and even presented
Louis-Napoléon with honors. The French persisted and began to
amass troops along the Swiss border. War was imminent. Then
Louis-Napoléon publicly declared that he would not be the cause
of war and voluntarily agreed to leave the country.

On October 25, 1838, the prince was back in London, having
been offered asylum in England. Louis Philippe's persecution of
the rebel prince greatly enhanced Bonaparte's reputation in England, and he was warmly welcomed by the international nobility living in London and by British high society. His new friends
included the Duke of Wellington, who had won the battle of
Waterloo, Benjamin Disraeli, and other members of Parliament.
Louis-Napoléon brought an entourage with him, to which he
added members from amongst his new and old friends in the
English capital. The prince spent his days pursuing the pleasures
of London and had a long succession of liaisons, even becoming
engaged once, but never marrying.

Louis-Napoléon joined a number of exclusive clubs. He

bought expensive clothes, practiced his equestrian skills, went on foxhunts with British nobility, and frequented the theater and the opera. He was comfortable with the life of the titled and wealthy, devoting himself to leisure, and almost forgetting his burning ambition to return France to Bonapartist rule. Then he perceived another opportunity.

On August 5, 1840, Louis-Napoléon hired the steamer *Edinburgh Castle* for what he called a pleasure cruise down the English Channel. The "cruise" included 113 passengers, and the crew must have been surprised to see that horses were brought on board, as well as strange large boxes with unknown contents. Even more surprisingly for a cruise, all the passengers were men, and they looked suspicious. At 2 A.M. on August 6, Louis-Napoléon summoned the bewildered captain, who now found himself surrounded by soldiers in uniform bearing arms. He was told to sail straight to the French city of Boulogne across the Channel.

The captain allowed his passengers to disembark outside Boulogne, and an hour before dawn, Louis-Napoléon and his little army advanced on a base of the French 42nd Regiment. The prince himself tried to rally the soldiers of the 42nd to join him in a march to Paris. "I make you an officer, now march with us to Paris," he said to a sleepy sergeant he woke up at the barracks.[43] Gradually, Louis-Napoléon and his men gathered the men at the base, and the prince addressed them all at dawn. He told them that a revolution was sweeping through France and that they should follow him. "Today, the great ghost of Napoléon addresses you through my voice," he announced. Before he was able to finish the speech, the commanding officer of the base,

Captain Col-Puygellier, who had recently woken up, came upon them on his horse. "Long live the King!" he shouted. Louis-Napoléon tried to offer him anything he wanted if he and his men would join the revolt, but the officer, loyal to his king, refused. "You would have to kill me," he told the prince, who would not do it. Seeing that he was not going to convince the troops to follow him, Louis-Napoléon marched his soldiers out of the barracks and into the town of Boulogne, which was slowly waking up to posters the rebels had plastered on the walls, urging the population to join the revolt to return France to its Napoléonic glory.

It soon became clear to the invaders that the citizens of Boulogne were not interested in revolt. The town had prospered under Louis Philippe, and no one felt an urge to overthrow the king. In disappointment, Louis-Napoléon retreated with his little army back to the port and attempted to board boats to return to the *Edinburgh Castle*. But as the small band floated on boats and buoys and anything they could find, they were pursued by the king's army. Louis-Napoléon was plucked out of the water and, with many of his men, taken to the local prison.

This time, the prince was not so lucky. When a few days later he was again brought before the King of the French, the latter was much less worried about reviving the Napoléonic mystique than he had been earlier. And he wanted to put an end to these attempts to replace him.

First, Louis-Napoléon was sent to the dreaded Conciergerie, on the Île de la Cité at the heart of the capital. This was an ancient palace dating to medieval times, named after the *concierge,* the king's keeper of the palace. It had long been a prison, and during the Revolution, half a century earlier, Marie Antoinette and many others had spent their days here before their executions. But Louis Philippe had no intention of executing this potential usurper. Louis-Napoléon was given the best cell in this high-vaulted, musty old citadel. Pityingly, the newspapers lambasted his folly. "This outdoes comedy," said the *Journal des Débats*. Other newspapers in France called Louis-Napoléon's attempted revolt "a miserable affair" and described it as grotesque and absurd. When he was called in front of his judges, Louis-Napoléon tried to raise the Napoléonic mystique and the ghost of Austerlitz, but it was no use. The prince was convicted and sentenced to perpetual imprisonment.

The king sent Louis-Napoléon to the medieval fortress of Ham, in Picardy, there to be held indefinitely. This was a foreboding old château, but again the prince had special quarters and many privileges, both because of his status as a prince and a Bonaparte and because of his money. Prisoners in France of those days could pay for better food and better conditions, if they could afford them. And Louis-Napoléon certainly could. He had a garden and relatively pleasant rooms and lived in reasonable comfort. He could interact with people and move around at will. He even had a little dog, whom he named Ham. Louis-Napoléon developed a relationship with a young woman who worked in the prison's laundry, Alexandrine Vergeot, and she bore him two sons.

While in prison, Louis-Napoléon taught himself science and carried out experiments, and so began a lifelong interest in the sciences. He would later refer to his years in the Ham fortress as time

The Prisoner of Ham: Louis-Napoléon performing scientific experiments in prison. *(Bibliothèque nationale de France)*

spent at "Ham University." Louis-Napoléon was especially interested in physics. He studied several areas of physics and even wrote a paper about electromagnetism. Years later, this deep interest in physics would attract him to another person who taught himself science, an unappreciated physicist who shared his ambition for success as well as the constant struggle to achieve it against heavy odds in a hostile environment. At the same time that, in Paris, Léon Foucault was photographing the Sun in early 1845 and conducting experiments with light and with microscopes, in the Ham prison a few dozen kilometers to the north, Louis-Napoléon Bonaparte was carrying out his own experiments with zinc, copper, and acid, and writing papers about these experiments. There was an invisible bond between these two self-taught scientists—one that would become real and strong within a few years.

• • •

On December 25, 1845, after five years in the fortress of Ham, Louis-Napoléon wrote a letter to the Minister of the Interior. In his polite, formal letter, he requested permission from the government to leave the prison temporarily for a trip to Italy to visit his ailing father. The minister replied that letting the prince leave the prison would be "contrary to the law" and declined the request. A friend, Sylvestre Poggioli, convinced Louis-Napoléon to write directly to the king. And so, on January 14, 1846, Louis-Napoléon wrote to Louis Philippe, King of the French:

> *Sire,*
>
> *I appeal to your sentiments of humanity . . . to allow me to visit my gravely ill father in Florence . . .*[44]

But the king turned down the appeal. Poggioli implored the prince to write again. But Louis-Napoléon declined. He had other plans in mind, and he had many friends who could help him.

If all doors for leaving the prison with permission were closed to him, Louis-Napoléon would get out of prison on his own. His friends had arranged a fake passport for him and helped him plan his escape. On May 25, 1846, Louis-Napoléon's valet, a man called Charles Thélin, entered the Ham prison dressed as a worker involved in some construction work inside the prison. He came to Louis-Napoléon's quarters and quickly shaved off the prince's beard and applied makeup to change his looks. He then dressed him in worker's clothes and had him carry a wide wooden plank to hide his face. Thélin then carried out a maneuver to divert the attention of the guard posted to the prince's quarters, walking the dog, Ham, in front of him, while the "worker" carrying the wooden board walked past right behind him and into the courtyard.

The guards carefully checked papers when the workers came in the morning and when they left in the evening, but not if workers left for a short work-related activity in the middle of the day. So Thélin and Louis-Napoléon walked out of prison in plain sight. At one point, they had to pass right under the balcony of the prison's governor. No one checked as the two "workers" exited the prison gate and crossed the bridge to freedom. A friend who remained in the prison covered Louis-Napoléon's bed with a blanket, having placed items on the bed that made it look as if the prisoner was sleeping. He told the guards that Louis-Napoléon wasn't feeling well and that they should let him sleep, then quickly exited the prison himself. The dog was left in good hands inside the prison.

With the help of the friend waiting outside, who quickly whisked them away, Louis-Napoléon and Thélin caught a train to Valenciennes, a city near the Belgian border. Before evening fell, they were safely across the border, and by the time the governor of the prison was informed that the prize prisoner of Ham had escaped, Louis-Napoléon and his cronies were celebrating the successful operation with a sumptuous dinner in Brussels. The next day, Louis-Napoléon reached Ostend, crossed the English Channel, and by evening he was back in London to a hero's welcome by his long-awaiting friends.

His daring escape and picaresque adventures made him a hero in Britain, and he was again attracting the attention, friendship, and admiration of many in the English capital. He once more resumed the life of a dandy, apparently forgetting—at least for the time being—his mission in life to recapture the glory of Napoléonic France. Again he began to pursue women and had a string of tempestuous relationships with theater actresses, aristocratic English ladies, and a courtesan by the name of Harriet Howard.

Miss Howard, as she was known, was born Elizabeth Ann Haryett in Brighton in 1823 to a family that made fashionable shoes and boots for the British aristocracy. She attended a prestigious school and also took dancing and riding lessons. Her goal was to become a Shakespearean actress, and she moved to London and assumed the name Harriet Howard. Miss Howard had great ambition but landed only minor roles in second-rate plays, for which she was hired more for her stunning good looks than for her acting skills. She was a better horsewoman than an actress and became famous for riding in parades. After a series of affairs, she met a major in the cavalry who had inherited a large amount of money. Since he was married, he installed her in a luxurious house in London, and after she bore him a son, he gave her a significant amount of money and property to ensure the son's prosperity.

Louis-Napoléon met Harriet Howard at Lady Blessington's house one night in the summer of 1846. The two became lovers, and their relationship was more constant than any of the prince's previous ones. He realized that money might hold the promise of his return to France. And her ambition drove her to solidify their relationship, as she saw in it a great opportunity to be France's first lady. Miss Howard encouraged Louis-Napoléon in his plans to return to France and readily offered him all the financial support within her means.

Bonaparte didn't have to wait long to return to his beloved France. Less than two years later, in 1848, revolution again swept through Europe. In France, Louis Philippe's "July Monarchy," as his reign was called, came to an abrupt end, and the nation inaugurated the Second Republic. This happened after a succession of events, beginning on February 22, 1848, when thousands of people streamed onto the streets of Paris to demonstrate for suffrage. A crowd gathered in front of the prime minister's office, and one of the guards shot at the crowd in an effort to disperse it. Seeing blood, the crowd became more enraged and the riot intensified. The king called in the National Guard, but this body, which had a history of anti-Royalist sympathies, refused to fire on their fellow Frenchmen. When he realized that the military would not defend him against his people, Louis Philippe fled to England with his wife, registering on board a cross-channel steamer as "Mr. and Mrs. Smith."

Ignoring the law that still forbade a Bonaparte to reside in France, Louis-Napoléon quickly boarded a steamship crossing the Channel in the other direction. From the coast of Normandy, Louis-Napoléon and his entourage took a train to Paris, arriving at the Gare du Nord, and then continued by carriage to the Hôtel des Princes on rue Richelieu.[45] The Bonapartists were rallying to his flag, posting his name all over Paris, but the new

French government was concerned and sent messengers to his hotel to remind him of the law forbidding him to reside in France. Fearing he might be forced to return to the Ham prison, Louis-Napoléon checked out of the hotel and left for England. Ironically, his distance from France worked in his favor, for the provisional government foundered, and the French people did not associate Bonaparte with this failure. On April 23, 1848, elections were held for the 880 seats of the National Assembly. Among the representatives elected that day were three relatives of Louis-Napoléon Bonaparte. There were also two Aragos: the father and son, François and Emmanuel.

François Arago was also invited to join the new French government that took over from the king and organized the Second Republic. He became the Minister of War as well as the Minister of the Navy. In his capacity, Arago abolished slavery in 1848 in the entire territory of France and its possessions, including Martinique, Guadeloupe, and French Guyana, where slavery had been common. France had thus abolished slavery fifteen years before the United States. Arago also championed the cause of universal suffrage in France.

In June, more elections were held, this time for eleven seats in the National Assembly, and Louis-Napoléon's name was placed on the ballot even though he was still in England. He prepared for this campaign and used Miss Howard's money, as well as his own, to print and distribute pamphlets and posters and to rally support for his election. Louis-Napoléon was elected, but the government immediately moved to declare his election invalid. The Bonapartists in the National Assembly united with the socialists and argued against the government, introducing a bill in the Assembly to censure the government. The debate spilled out of the National Assembly. In the streets of Paris, the name Louis-Napoléon Bonaparte was on everyone's lips. The prince was quickly becoming the most famous man in France. He added to his popularity by volunteering to withdraw from the seat he had just won in the National Assembly to preserve the unity of France. In the meantime, he raised more money for his campaign by liquidating some of his properties in Italy and properties belonging to his mistress. In September, when elections for the National Assembly seats were again called, Louis-Napoléon received an overwhelming number of votes: over 110,000. He immediately left for Paris. Miss Howard soon followed him there and settled in a hotel a discreet distance from his.

On September 26, 1848, Louis-Napoléon took his oath of office and became a member of the National Assembly. He performed his new civic duty with little distinction, and few people considered him a good orator. But he had an aura. As historians describing the period would later say, his name spoke for him. On December 10, 1848, the French held a national election for President of the Republic. France was headed at the time by General Eugene Cavaignac, a man who as head of the military had brutally quelled a rebellion by the masses some time earlier. In doing so, Cavaignac had antagonized the working classes, but was supported by right-wing elements in the National Assembly.

Cavaignac's supporters attempted complicated legislative maneuvers to keep Louis-Napoléon from being elected President. But Bonaparte's name worked for him, and he also had money to support a strong presidential campaign. When the results of the election came in ten days later, Louis-Napoléon had 5.4 million votes to Cavaignac's 1.4 million. The Napoléonic mystique had prevailed. Many among the masses in France pined for the old empire, and even though he had died in Saint Helena years earlier, many in the lower classes believed that the first Napoléon was still alive. So support for the nephew now running for president was strong. On December 20, 1848, Louis-Napoléon was sworn in as President of the Republic.

Once again a Bonaparte was at the head of the French nation. He'd had an amazing ride—two attempts to take power from the monarchy, six years in prison, a daring escape, years of exile abroad—but now Louis-Napoléon Bonaparte was President of the French Republic. And he had many plans for the nation. He held progressive views, and he dreamed of building France into a modern state with a good banking system, an extensive railroad network, and a thriving economy. He also liked to enjoy his new life as head of the nation.

Instead of spending hours reading reports by committees and bills presented to the National Assembly, the Prince-President reportedly would wake up around 10 A.M. He would spend two hours dressing and having breakfast, and then spend about two hours doing state business. At 1 P.M. he would hold a cabinet meeting, lasting until 3:00. Then Louis-Napoléon would take a carriage to the Rond Point des Champs-Elysées. This was the location of his rendezvous with Miss Howard. She would get there in her own carriage from her house on the rue du Cirque. The two would mount horses and ride together for two hours in the Bois de Boulogne. After that they would have drinks at a café. In the evenings, he would often come to her house.[46] This was more the life of a king than a president, and Louis-Napoléon clearly desired to rule as a monarch. Miss Howard sensed it and encouraged it. She wanted to be the wife of the absolute ruler of a nation. She craved the luxury of royalty.

Louis-Napoléon came to his job having trained himself in many fields: politics, economics, and science. This last area would be important to him throughout his life. He followed what took place at the meetings of the Academy of Sciences, would read the *Proceedings,* and discuss issues with scientists. The Academy served an important role for the government. Members of the Academy would advise the President on how to best resite hospitals, plan a railroad system, immunize the population against disease, treat drinking water, and more. And through his work with the National Assembly, Louis-Napoléon became acquainted with François Arago.

So it was by this intimate connection with the scientific community and the Academy of Sciences that in February 1851 Louis-Napoléon Bonaparte, the Prince-President of the Republic, found out about the experiment that the leading scientists in the French capital had just witnessed in Meridian Hall of the Paris Observatory.

Louis-Napoléon's reaction was immediate. He decreed, as President of the Republic, that the Foucault pendulum experiment be repeated in the highest and most famous dome in all of Paris, inside a building that had come to symbolize everything the French nation held sacred: the Panthéon. And he added to his presidential order that "this should be done with the speed of lightning."

9

◆

THE FORCE OF CORIOLIS

When Louis-Napoléon's order was pronounced, in early 1851, the scientific community and the Academy of Sciences were in the midst of attempts to grapple with what Foucault had proved and to deal with the embarrassment dealt them by his achievement of providing the sine law without using equations or mathematical derivations.

In response to Foucault's proof, the mathematicians searched every theoretical paper they could find to try to explain the phenomenon of Foucault's pendulum and cited a slew of theoretical results. In addition to work by the mathematicians mentioned earlier, Jacques Gapaillard cites in his book *Et Pourtant Elle Tourne* works by the great Gauss, Hamilton, Jacobi, Laplace, and d'Alembert.[47] It seems that every important mathematician of the seventeenth, eighteenth, and nineteenth centuries had published work that—after the fact of Foucault's experiment and his sine law—was seen to have something to say about the subject or some way of explaining why the pendulum reacts the way it does to the rotation of the Earth. Gapaillard even goes on to suggest that Foucault's work caused a rift between two camps within the Academy of Sciences: the analysts and the synthesists. One group believed only in pure mathematics and the geometry of the pendulum experiment, while the other attempted to explain things in terms of the kinematics of moving objects, that is, the physics of mechanics, rather than geometry.

But in their race to interpret Foucault's discovery, both groups of scholars missed an explanation of Foucault's pendulum within the work of a little-known French physicist by the name of Gaspard-Gustave Coriolis (1792–1843), who had died just eight years earlier.

Gaspard-Gustave Coriolis was born in Paris at the time of the French Revolution. His father, Jean-Baptiste-Elzéar Coriolis had served in the Bourbonnais regiment, which fought in the American campaign in 1780. He returned to France and was promoted to captain. He then became an officer serving Louis XVI, which jeopardized his safety when the monarchy was toppled. The king tried to escape the mobs and fled Paris on June 21, 1791, but was caught at Varennes and brought back to the capital, where he was eventually guillotined. Gaspard-Gustave Coriolis was born in June 1792, just as the monarchy was abolished.

Coriolis was brought up in Nancy, where his family had settled after the Revolution. In 1808 he took the entrance examination for the École Polytechnique and was placed second of all the students entering that year. After graduation, he entered the École des Ponts et Chaussées (the School of Bridges and Roads) in Paris. Following his father's death, Coriolis had to support the family. He suffered from poor health, which didn't leave him many possibilities for work. He chose to take an appointment as a tutor in mathematics at the École Polytechnique, and Cauchy recommended him for the position. After the July 1830 revolution that brought Louis Philippe to the throne, Cauchy left Paris, refusing to swear an oath of allegiance to the new king. The École Polytechnique then offered Coriolis the position that had been held by Cauchy, but he declined, since he wanted to limit his involvement in teaching. In 1836, he was appointed chair at the École des Ponts and Chaussées and was also elected to the mechanics section of the Academy of Sciences. But his poor health deteriorated further in the spring of 1843, and he died a few months later.

Coriolis conducted research in mechanics and engineering. He studied friction, hydraulics, and the performance of machines. He introduced the physical terms *work* and *kinetic energy*. In 1835, Coriolis published a paper with the title "On the Equations of Relative Motion of Systems of Bodies." In this paper, Coriolis described what is now known as the *force of Coriolis*. The force of Coriolis is an unusual physical phenomenon that affects rotating systems.

• • •

Foucault, who began his scientific career working for Donné after the death of Coriolis, does not seem to have known anything about the work of Coriolis. Foucault, in all his explanations of the pendulum experiment, never mentioned Coriolis. And yet, it so happens that the best twenty-first-century explanations of the movement of the plane of oscillation of Foucault's pendulum are the ones that use the force of Coriolis. It is also a mystery that the members of the Academy of Sciences—of which Coriolis had been a member—never mentioned his work in the context of Foucault's pendulum.

The force of Coriolis is a "fictitious" force affecting bodies in rotation. It is thus similar to another "fictitious" force related to rotating bodies: the centrifugal force. Both are called "fictitious" because a body affected by these forces is not actually pushed by a material thing. But a person on a fast merry-go-round feels the centrifugal force as if it were a real force.

The force of Coriolis also affects a person on a merry-go-round, but only if that person moves. And it is a very unusual force. It acts on a person or body in a direction *perpendicular* to the direction of motion. That is, the Coriolis force acts *sideways*. And it feels quite bizarre. If you are on a merry-go-round, or a similar spinning surface, and you move inward, toward the center of the wheel, then you will feel a strange push to your right if the wheel spins counterclockwise. If the wheel spins clockwise, when you try to move closer in, you will feel a push to your left. If you move outward on a counterclockwise-moving merry-go-round, you will feel a push to your right. And the opposite will happen (a force pushing you to the left) if you move outward on a clockwise-turning wheel. Notice that everyone, even at rest, when placed on any spinning wheel, will feel a force pushing the person *outward,* regardless of the direction of spin (clockwise or counterclockwise): This is the centrifugal force in action.

So what causes this weird Coriolis force, which affects only moving bodies on a spinning surface, never stationary ones on the rotating surface, and is perpendicular to the motion on the surface?

There is a nice explanation of the Coriolis force. Look at the figure below.

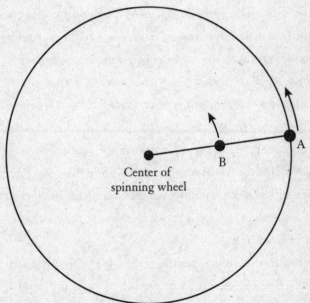

Center of
spinning wheel

Imagine that you are standing at point A, at the edge of the wheel spinning counterclockwise (the direction of rotation of the Earth as seen from above the north pole). Now, the wheel is spinning at a given angular velocity (number of revolutions per unit of time; say it is spinning at ten revolutions per minute). This rotational speed gives you some specific *actual* speed (say, ten miles an hour). Now suppose you want to walk over to point B, which is farther *inward* on the wheel from where you are. But at the same angular velocity of the wheel (A and B are on the same spinning surface) point B, being closer to the center, moves around at a smaller *actual* speed (say, only 7 miles per hour; remember, the center of the wheel, while spinning, has zero actual speed, since it is not moving, only spinning around). So as you are moving inward toward point B, you are moving to a point that moves around at a slower speed than you are. Approaching point B, which moves slower than you, you "feel" your greater speed around as a force pushing you sideways away from B. This is the force of Coriolis.

The force of Coriolis is responsible for the direction of spin of storms, hurricanes, and tornados in the northern hemisphere, which is opposite of that of storms in the southern hemisphere. The Coriolis force is also responsible for the theoretical direction of the spin of water going down the drain (counterclockwise) in the northern hemisphere versus the spin in the southern hemisphere (clockwise).

But this rarely happens in reality. If the sink has some imperfections that favor the water going one way or the other, that effect will dominate the Coriolis effect. The water in the sink tends to choose its own direction of spin, regardless of the Coriolis effect, which is very weak on this scale of magnitude. There are stories about fraudulent tourist traps at locations on the equator where a local person charges visitors for a demonstration in which north of an imaginary line the charlatan claims is exactly the equator water drains in one direction, and right below it, in the opposite direction. By a slight spin given to the water, a sleight of hand, this person can make the water entering the drain spin around in any direction he or she chooses. The Coriolis effect is subtle and can be disturbed by outside influence.

But the Coriolis effect is real enough. As mentioned earlier, artillery engineers have always known that (in the northern hemisphere) cannonballs fired north tended to deviate east, and ones fired south tended to veer west of their destination. These deviations can be explained by the force of Coriolis. The situation is similar to that of a child on a merry-go-round. When a cannon fires north from a location in the northern hemisphere, the location of landing is closer to the pole and hence moves at a slower speed than does a point on the more southern latitude from which the cannon was fired. The cannonball transports with it its greater radial speed (eastward, with the rotation of the Earth) and thus it lands farther east than expected.

The eastern deviation that Newton had predicted for falling objects, which was studied by Hooke, Gauss, Guglielmini, Benzenberg, and Reich, can also be explained by the force of Coriolis. On the top of a tower, the body to be dropped down to Earth has a greater radial velocity than does a point on the ground. This is true because a point on the ground is closer to the center of the Earth, and thus moves around (eastward) more slowly than does a point on the top of the tower, since it is located farther from the center of the Earth. When the object falls from the top of the tower to the ground below, it carries with it its greater easterly velocity and hence lands to the east of the location we expect it to fall to.

Let's apply the discussion of the Coriolis effect on a spinning wheel to a body or person on the spinning Earth, using actual numbers.

The Earth completes a rotation in twenty-four hours. The equator is about 25,000 miles in circumference. So if you are standing on the equator you are moving in space at a speed of 1,042 miles per hour due to the rotation of the Earth. In southern Florida, at 27 degrees north, you are moving at a speed of only 930 miles per hour, and in Barrow, Alaska, at 71 degrees north, your speed due to this rotation is only 340 miles per hour. And exactly at the north pole your speed is zero. An airplane, flying directly from southern Florida to Barrow, Alaska, would therefore have to adjust its flight as it will have an added speed of 930-340=590 miles to the east as it goes toward Barrow. A cannonball fired from one place to the other would be deflected a long way from its destination because of this added speed due to the rotation of the Earth.

If we understand the Coriolis effect, it is also easy to see how storms develop in the two hemispheres. In the northern hemisphere, when air rushes south from a high-pressure area to fill a low-pressure zone, the Coriolis effect will deflect it west. At the same time, air flowing north to fill the same depression will be deflected east by the Coriolis effect. This creates a storm that rotates counterclockwise. In the southern hemisphere the situation is reversed. See the figure below.

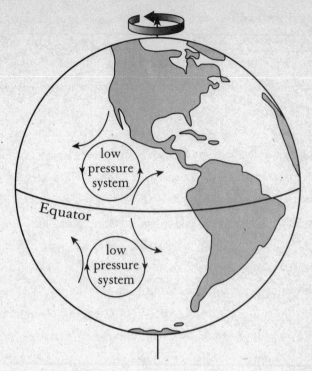

The Coriolis effect can be used to explain what happens to the Foucault pendulum.

As the pendulum swings north, the force of Coriolis pushes it ever so slightly east (in the northern hemisphere, say in Paris). When it swings south, the Coriolis force sends it slightly to the west. As this motion repeats itself over a long period of time, we get the rotation of the plane of oscillation around a circle on the ground. To see what happens when the pendulum starts its swing on a perfectly east-west plane requires deeper mathematical analysis.

But Foucault needed neither the force of Coriolis nor the equations of motion nor the rigorous geometrical derivations of the mathematicians to explain the motion of his pendulum. According to Joseph Bertrand, a member of the Academy of Sciences who wrote an article about Foucault in the annals of the Institut de France in 1882, Foucault had an ingenious method all his own.

Foucault carried around with him a small wooden ball representing the Earth. This little ball was covered with notations and figures, and according to Bertrand, through the study of this model Foucault was able to derive his sine law. Foucault loved to walk in the Luxembourg Garden, contemplating his wooden ball. One day, according to Bertrand, Foucault met a friend—a mathematician—and asked him to use his equations to calculate a small angle pertinent to the swing of the pendulum. The friend sat down on the grass and carried out a lengthy computation, while Foucault just stood by with his fingers on a triangle on his little wooden ball. Finally, the friend looked up and gave the answer. "I thought so," said Foucault, pointing to the triangle on his little ball, which gave him the answer directly. He walked away, "a triumphant smile lighting his frail features."[48]

10

———◆———

THE PANTHÉON

The Panthéon, a large neoclassical building with a high dome, dominates the Paris skyline. This is the third point of our imaginary triangle mentioned in the Preface encompassing an important part of the Left Bank of Paris. Leaving the magnificent Luxembourg Garden through its eastern exit by Boulevard Saint-Michel, one sees the high dome of the Panthéon rising up on the hill at the top of rue Soufflot, named after the builder of the edifice. When it is flooded with white light at night, the Panthéon makes an even more impressive sight. It looms over its surroundings, and, even from a great distance, it appears to be nearby.

In the 1850s the area around the Panthéon, the Mont Sainte Geneviève, held all of the universities of Paris. Right next to the Panthéon is the rue de la Sorbonne and, just below, the Place de la Sorbonne, the seat of the ancient University of Paris. The other schools of the capital were at that time (some have moved since then) arranged along the streets that emanate from the Panthéon as if it, too, were an "étoile" (a star), as the collection of intersections around the Arch of Triumph is called today. The streets around the Panthéon all extend outwards from its high location at the top of the ancient hill, Mont Sainte Geneviève. South and east of the Panthéon, on the rue d'Ulm and the rue Lhomond, are the buildings of the École Normale Supèrieure; and to the north, on rue Descartes, was the École Polytechnique; while just down the hill, to the north, is the rue des Écoles and by it stands the Collège de France. Because of the concentration of schools and universities, this area around the Panthéon has been called the Latin Quarter, for the students and professors in the sixteenth and seventeenth centuries spoke Latin at the university.

The Panthéon was the perfect site for Foucault's demonstration of the rotation of the Earth because of the building's importance, and because the high dome would make the experience magnificent. But the Prince-President may have also wanted to send a message to the academics of the Latin Quarter—for members of the Academy of Sciences held teaching and research positions at the universities on this hill, and none of them could ignore the Panthéon in their midst and the historic event about to take place inside it.

Like the Île de la Cité, the heart of Roman Paris called Lutetia, the hill across the river and south of that island has been inhabited since early times. In 451 A.D., Attila the Hun was preparing to attack Lutetia. A young shepherdess named Geneviève took the lead and organized the people of the surrounding area to resist Attila's armies. The Huns, for whatever reason, never did attack. To commemorate their deliverance, the people buried Geneviève on top of the hill that today bears her name when she died in 502 A.D. Later she would become a saint, the patron saint of Paris.

In 508 A.D., Clovis, who was France's first Christian king, replaced the modest shrine on top of the mountain commemorating Geneviève with a church dedicated to the saints Peter and Paul. When he died, Clovis himself was buried at the site and so was his wife, Clothilde, and thus began the custom of burying famous or important people at this site—a tradition that would continue and increase in frequency in the modern French state. For the Panthéon would serve at times as a church and at other times as a secular mausoleum to France's greatest men and women.

In the ninth century a new and larger church was built at the same place on top of the hill, and it was renamed after Geneviève. The church was enlarged in the twelfth century and, thereafter, Sainte Geneviève was revered in Paris and her relics carried in procession whenever the city was in danger of attack or disease or flood.

In 1744, Louis XV fell ill and prayed to Sainte Geneviève. When he recovered, he came to the church of the patron saint and was dismayed at its neglected state. The king made a promise to rebuild the church, and funding was finally arranged ten years later for the project to begin. The royal involvement in the rebuilding of the church underscored the alliance of the French monarchy with the Church.

The commission to build a large and impressive church on the site of the old one was given to the architect Jacques-Germain Soufflot (1713–1780), who was known for his innovative ideas. He had made many trips to Italy, and his style was influenced by Italian design. He wanted his Church of Sainte Geneviève to rival the great Basilica of Saint Peter in Rome.

Soufflot's aim was thus to incorporate elements of both Gothic and Greek architecture. He chose weight-bearing columns in the Antique style, and he used the design for great height and high luminosity typical of the Gothic style. The first stage of construction was finished in 1755. It consisted of a Greek cross plan with four naves of equal length. Above the crossing area rose a high dome. The design was modified five times over the years, and the final version of the design was finished in 1777. Soufflot used different kinds of vaults: cupolas and barrel vaults, which concentrated the pressure at specific points, while the sides of the building were supported by flying buttresses. The dome was so immense that none of the supports alone could bear its weight, and so Soufflot constructed four pillars for the purpose of holding it up.

This Herculean construction project attracted early criticism. A competing architect complained in 1770 that the huge dome could not possibly be stable. Soufflot anticipated all criticism by using an iron framework designed to help support the dome. The final result of Soufflot's plan and eventual construction was a truly massive edifice. As you walk around this curiously cross-shaped, enormous building, it seems to rise straight up into the air on all its sides. At its front, facing what is now rue Soufflot and the Luxembourg Garden, colossal columns greet the visitor. And the dome itself is supported by another set of smaller columns. In addition to Saint Peter's Cathedral in Rome, it also resembles its namesake, the Roman Pantheon, built seventeen centuries earlier. It has the same Roman massiveness.

Louis XV laid the cornerstone for the Church of Sainte Geneviève in a ceremony on September 6, 1764, but the construction work continued until 1790, ten years after Soufflot's death. Soufflot himself had planned how to decorate the inside of the church. The church was to be filled with many beautiful works, which later would include paintings by Antoine-Jean Gros (1771–1835), who painted a fresco on the cupola. But before this was done, just as the construction was nearing completion, the Revolution erupted.

The revolutionaries had different ideas about this building. The Marquis de Vilette wanted the church to be a resting place for his friend Voltaire. And thus the idea for the Panthéon was born: a place in which France could bury those who contributed to its national glory. Two years later, the National Assembly was petitioned to make the Church of Sainte Geneviève "a temple of the nation, an altar to liberty."

The idea was taken up with enthusiasm, and following Voltaire, Marat and Rousseau were also "Panthéonized." The custom continued over the years and is retained to this day. Pierre and Marie Curie were also buried in the Panthéon, and so were Balzac, Victor Hugo, Émile Zola, André Malraux, and Antoine de Saint-Exupéry.

But the switch from church to secular temple and back to church would continue. In 1806, Napoléon I, having signed the Concordat with the Pope a few years earlier, returned the Panthéon to its former function as a church, while the crypt retained its civic character as the final resting place for the nation's greatest citizens. In 1811, Napoléon commissioned the artist Gros to paint the cupola. Gros painted angels carrying the relics of Sainte Geneviève to heaven, and below it France's famous kings, Clovis, Charlemagne, and Saint Louis. And he drew his patrons, the emperor and the empress.

When the monarchy was reinstated after Napoléon's defeat, the Panthéon again resumed its function as a church. Then, in 1830, the building became secular again when the July Monarchy of Louis Philippe took over from Charles X and the Bourbons. But out of political considerations, Louis Philippe later closed the building altogether, and there were no Panthéonizations until the end of his reign. The work of incorporating new art continued inside.

When the 1848 revolution replaced the reign of Louis Philippe with the Second Republic, the Panthéon was renamed "The Temple of Humanity." It was to this newly designated secular temple that Foucault would bring his pendulum.

Working very quickly, as the Prince-President had ordered, Foucault again hired Froment to create a new pendulum. The craftsman made his best pendulum, itself a work of art (now displayed

The inside of the Panthéon (with a replica of Foucault's pendulum).
(Corbis-Bettmann)

at the museum of the Conservatoire National des Arts et Métiers in Paris). The bob of the pendulum was made of brass and weighed 28 kilograms. It was 38 centimeters in diameter. Since the dome of the Panthéon is so high, Foucault had a wire 67 meters long made for this pendulum. This was undoubtedly the longest and heaviest pendulum the world had ever seen. Foucault must have smiled while preparing for his presidentially decreed public demonstration. This would be a magnificent spectacle.

The original bob of Foucault's pendulum, made by Froment, used in the 1851 demonstration in the Panthéon. (*CNAM, Paris*)

But what was in it for the Prince-President? Why was Louis-Napoléon so interested in Foucault and his pendulum? Louis-Napoléon certainly had a deep interest in science. He had studied the sciences during those lonely months and years at the Ham prison, and he continued to pursue his interests in science, technology, and progress as the President of the Republic. We know that Louis-Napoléon was a Saint-Simonist, a follower of the philosophy of Claude Saint-Simon (1760–1825). This theory held that the key to progress lay in socialism linked with science and technology. For practical reasons, as well, Louis-Napoléon was firmly convinced that following scientific ideas as they developed was important for the modern state. His predecessors in power had consulted with the Academy of Sciences on scientific issues that affected society and its technological advancement. We know from Louis-Napoléon's legacy that he always followed a path to progress through innovative ideas of science. In this sense, Louis-Napoléon was a modern ruler, despite the terrible events that would transpire within a few months.

We also know that Louis-Napoléon sincerely liked Foucault—perhaps he even identified with him in some way. Both men were extremely capable individuals, and both had been, or at least had felt, unappreciated by society. Both had a burning desire to prove their talents to the world. This was Louis-Napoléon's chance to help a struggling genius assert his talents despite opposition from the established science community. And these other scientists were no friends of this Bonaparte, despite the relationship that his illustrious uncle had with the two mathematicians Monge and Fourier. In particular, Arago—this immensely influential scientific Brahmin and legislator—was no friend to the Prince-President. The entire Arago family had been staunchly republican in their political

views. They considered the election of Louis-Napoléon Bonaparte to be President of the Republic on the 10th of December
1848 as the end of the new Republic, which had just done away
with the rule of the last king, Louis Philippe. Jacques Arago, a
first cousin of François, reported in his memoirs a remarkable
dialog that supposedly took place between his brother Étienne
and Louis-Napoléon after the latter had declared his candidacy
for President. Étienne said to the prince: "You will be forced to
march toward the monarchy. What will push you forward is
the ignorance of the masses and imperial fetishism!" Louis-
Napoléon supposedly invoked the memory of his uncle's service
to the republic, to which Étienne Arago replied by reminding
him that his uncle did, however, become emperor. He added,
"This is why, Sir, an enlightened and sincere republican will
never vote for you."

When the election itself took place, François Arago himself—who was not a candidate in the election—received in his
region of the Eastern Pyrenees 441 votes in his name, against 82
for Louis-Napoléon. The confrontation between the Arago family and Louis-Napoléon Bonaparte would continue.

In the National Assembly, Louis-Napoléon did not work effectively. He had problems dealing with the democratic process.
The *Journal des Débats* reported on January 4, 1851, on the delib-
erations at the National Assembly of the day before. The President of the Republic was given ample opportunity to speak at
length about the issues at hand. Every now and then there were
laughs reported from the Right. At the end of the discussion of
some "closure" that was debated that day (whose details are

unimportant for our story), the following exchange was reported to have taken place:

> *Louis-Napoléon Bonaparte: I have only one word to say . . .*
> *Voice: The closure! It is the order of the day pure and simple.*
> *Louis-Napoléon Bonaparte: My question . . .*
> *The President of the Assembly: The Assembly announces with immense majority for closure.*
> *Louis-Napoléon Bonaparte: I demand the word against closure.*
> *The President of the Assembly: Closure is pronounced.*
> *Louis-Napoléon Bonaparte: I've demanded the word against closure . . .*
> *The President of the Assembly: The Assembly did not want to listen to you.*
> *Louis-Napoléon Bonaparte: It doesn't have the right to violate its own regulations by closure.*
> *The President of the Assembly: I consult the Assembly on the order of the day pure and simple.*
> *Louis-Napoléon Bonaparte: I take back the order of the day because . . .*
> *Voices: Too late. Too late!*
> *The President of the Assembly: You cannot prevent the Assembly from doing its job. The order of the day pure and simple is pronounced.*[49]

It is not hard to see why a President who had such difficulties working with his nation's elected representatives would cherish the decrees he could make on his own, regardless of what the legislators might want to do. Decreeing that a public demonstration of the rotation of the Earth by way of a pendulum experiment be conducted was just one such example. Foucault's contemporary biographer and close friend, Philippe-Louis Gilbert, said in his manuscript, *Léon Foucault, His Life and Scientific Work,* published in 1879, that Louis-Napoléon always liked Foucault, held him in high regard as a scientist, and kept up with his work and achievements.[50] All of these were good reasons for the Prince-President to decree that Foucault should have the great Pan-théon, the *Temple of the Nation,* at his disposal. The President of the Republic could do so without consulting the Assembly, with its voices against him, and despite people such as Arago, who was not only a venerated scientist and Perpetual Secretary of the Academy of Sciences—of which Foucault was not a member—but also a legislator with often opposing agendas. He must have felt there was some poetic justice in his promoting the work of Foucault in such a public way despite what the scientists might think.

Finally, there was one more, obvious reason. The monar-chies have always aligned themselves with the Church. This al-lowed them to perpetuate the myth that the King was sanctioned by God. The republic favored secular power and turned the great Church of Sainte Geneviève into the Panthéon. Louis-Napoléon's order that Foucault's pendulum—providing evidence against the biblical view of a stationary Earth—be demonstrated in the Panthéon was a nod at secular power and against both the monarchy and the Church. Louis-Napoléon was testing his powers. Who would he need most on his side:

scientists, the Church, the masses, intellectuals? The public spectacle of the pendulum experiment could provide him with some clues.

If the scientists lost interest in Foucault and his work after the initial excitement over his experiment in Meridian Hall had faded, the public in general began to notice him. Foucault was, in a sense, the scientist of the people. He was one of them, not a prestigious, qualified scientist. Already on March 19, 1851, the science reporter of the newpaper *Le National,* a person named M. Terrien, reported on the Meridian Hall experiment and added:

> M. Foucault will not stop here, I hope. An experiment of this quality could be carried out in public, on a grander scale and with a certain degree of ceremony. It is from the Panthéon dome that a wire some sixty meters long should be hung. Those noble walls, which lie bare and unproductive . . . are worthy of serving as a setting for one of the finest experiments ever invented in a physicist's mind.[51]

Foucault, of course, and his engineer, Froment, were at that time working hard to implement this very experiment, commanded by the President of the Republic. They and their workers climbed to the top of the high dome and installed the mechanism that would allow this huge pendulum to swing freely in any direction and change its plane of oscillation without any force preventing the movement.

On the floor of the Panthéon, centered under the dome, there was, and still is, an elegant design in marble. Foucault covered it with a wooden circle, six meters in diameter, over which his heavy pendulum would swing. Around the circle he erected a mahogany balustrade. The edge of the wooden circle was divided into degrees and quarters of a degree, so that the public could clearly see the movement of the pendulum's swing pattern over time. Finally, Foucault and Froment had the wooden disk covered with a layer of wet sand, just high enough so that the tracks of the pendulum as it swung would remain visible to observers. The two had all the help they would need—the President had issued specific orders down to the lowest levels of the administration to ensure that the project had all the support of the French Government.

Toward the end of March, newspapers in Paris announced the upcoming public display of Foucault's pendulum in the Panthéon, saying that the president, demonstrating his "support for science," has decided that the experiment should be conducted in a grand public forum.

On March 26, 1851, the science reporter Terrien again wrote an article in *Le National:*

Have you seen the Earth go round? Would you like to see it rotate? Go to the Panthéon on Thursday, and, until further notice, every following Thursday, from ten A.M. until noon. The remarkable experiment devised by M. Léon Foucault is carried out there, in the presence of the public, under the finest conditions in the world. And the pendu-

lum suspended by M. Froment's expert hand from Soufflot's dome clearly reveals to all eyes the movement of rotation of the Earth.[52]

The Parisians flocked to the nation's shrine to see the great
experiment in physics. This would be a historical moment, in
which Galileo, Bruno, and, of course, Copernicus and Kepler,
would be vindicated—inside a magnificent church now dedicated to greatness in science and letters and politics.

The Prince-President was there, along with dukes and duchesses
and counts and countesses, leaders of industry and business, and
average citizens. A smiling Foucault stood by the wooden circle,
waiting. Froment was standing to one side under the great
dome, from which the steel wire hung down holding the large
pendulum bob. The bob itself was secured by a strong thread to a
post on the side, near Froment. When Foucault issued the order,
Froment touched a lit match (safety matches had just been invented) to the thread. As it caught fire, the thread released the
pendulum, which swung down to the center of the circle and on
to the opposite side. People were watching the swinging pendulum with great interest and curiosity. As Foucault later described
it in an article in the newspaper *Journal des Débats*:

After a double oscillation lasting sixteen seconds, we saw it
return approximately 2.5 millimeters to the left of its starting point. As the same effect continued to take place with

each new oscillation of the pendulum, this deviation increased continuously, in proportion to the passing of time.[53]

The steel stylus that Froment had attached to the bottom of
the bob made a line in the sand as the pendulum passed over it.
These lines formed long and narrow ellipses in the sand, and the
people all around the wooden circle and mahogany balustrade
could clearly see how the tracks of the pendulum were veering
in a clockwise fashion. The public was fascinated—taken in by
science in the making. According to an eyewitness, every day
there were many people inside the Panthéon, looking at the
strange pendulum hanging from the high dome above—even
during times the pendulum was motionless and the experiment
not in action. Foucault himself was there for hours every day,
explaining to those around him what the pendulum was doing
and how it demonstrated that, in fact, it was the Earth itself that
was rotating under the pendulum. The plane of the swing of the

The Panthéon demonstration as seen through the eyes of a contemporary
artist. *(Académie des Sciences, Paris)*

pendulum was actually "fixed in absolute space," as he put it, "while we and the planet rotated right under it."

Foucault was a celebrity. But the world of science continued to ignore him. According to his biographer Stéphane Deligeorges, the attention given to Foucault by Louis-Napoléon made it worse for him, because it made the scientists and members of the Academy jealous. Not only was he not one of them—not a trained scientist and not a mathematician—but now he was getting all the media attention and was becoming the darling of Paris high society and the President of the Republic.

But Foucault was clearly enjoying his new status. The perennial bachelor was especially eager to explain his pendulum experiment to well-dressed, beautiful women who came to the Panthéon for the weekly demonstrations of the rotation of the Earth.

A short time later that year, Prince Louis-Napoléon Bonaparte bestowed on Foucault one of the greatest honors the French nation accords its heroes: Foucault became a member of the Legion of Honor. But he was still not a member of the Academy of Sciences—and thus not recognized by the scientific community as a peer—and he would not be granted this status for many years to come.

Foucault was not just a folk hero. He was, in fact, a great scientist. And he proved it—in typical fashion, since he was now recognized for his achievements by the general public—in a newspaper article. Foucault, who had been reporting for years in the *Journal des Débats* about the science of others, wrote on March 31, 1851, an article about his own work: his experiments with the pendulum.

It is rare in the history of science for a scientist to expose the details of a theory in a public forum rather than in a professional, scientific journal. One is reminded of Galileo, who in the 1600s chose to write about science in the vernacular—the Italian language of the masses—rather than in the customary Latin, the language of the savants.

At any rate, Léon Foucault's article, on page three of the *Journal des Débats* on Monday, March 31, 1851, is a historical document. Foucault described the phenomenon of the movement of the plane of oscillation of the pendulum in such a perfect, theoretically precise—and yet readily understandable—way that it must have put all the mathematicians to shame once again. His language and his use of examples showed that he had not only mastered the science, but was also capable of explaining it in simple terms that everyone could understand. The following is a translation of a part of Foucault's long article.

Before turning our attention to the movement of the Earth, which because of its slowness requires us to turn to well-developed methods, let's first install ourselves next to a table, which we can move at will, above which we place a small pendulum, that is, a lead ball suspended on a wire. The room in which we operate will be for us the universe; the table will represent the Earth. The pendulum, suspended from its support, will move above a circle, which will be traversed by several diameters, and their point of intersection will correspond to the direction of the pendulum while at rest. The pendulum, its support, and the circle will be viewed as a single apparatus, which we will place at the center of the table. We release the ball in the direction of one of the diameters on the circle. What will happen then? The simplest and most evident thing in the world: As soon as it is free, the pendulum launches itself toward the center of the circle, continues past it because of its speed, comes back, and swings back and forth until it stops at the center of the circle. Its plane of oscillation is constant in the direction of the diameter with which it was originally aligned. If we look for the reference frame for this movement outside the table, on the walls of the room, we reach the same conclusion. But if, while the pendulum is moving, we gently turn the table around, without jolting it, what would be the relation between the table and the plane of oscillation of the pendulum? Those of you who have not yet carried out this experiment, what would be your response to this question? Doesn't it seem to you, at first glance, that the plane of oscillation should turn with the table, so that the pendulum would continue along the same diameter drawn through the circle? Profound error! This is exactly the opposite of what will happen. The plane of oscillation of the pendulum is not a material object. It does not belong to the support, or to the table, or to the circle. It belongs to space—to absolute space.[54]

One of the most important ideas in physics is the notion that there are various *frames of reference*. The way a situation looks to one observer may not be the same as the way it looks through the eyes of another observer, who may be in motion relative to the first observer. This idea was first developed by Galileo in the 1600s and eventually extended by Einstein in 1905 to form the basic element of his special theory of relativity. In his newspaper article aimed at the general public, Foucault explained the motion of his pendulum in clear terms along the lines of this "Galilean relativity." His addressing the phenomenon of the pendulum in terms of various frames of reference shows a great deal of physical insight—of the kind possessed by Galileo and Newton before him and Einstein half a century after him. Modestly, Foucault ended his article with a nod at the mathematicians.

The analytical work of M. Binet, the lively remarks of M. Liouville, and the appreciated ingenuity of M. Poinsot bestow upon us the duty to continue our work with perseverance. We assure you we will not lack there.

The mathematicians were not impressed with this gesture. But the public was enthralled.

Over the next weeks and months, both Foucault and the science establishment—the Academy of Sciences, the Institut de France, and the schools and universities in Paris—received hundreds of letters from members of the public asking questions about the amazing experience of Foucault's pendulum. One letter, mentioned by a contemporary biographer, was sent to Foucault from a man who wrote:

Sir,

I shall be desirous to have one of your pendulums that marches to the movement of the Earth. Where might I procure one? I shall be most grateful to you if you should give me an indication.[55]

The public display of the pendulum in the Panthéon and the resulting publicity took the proof of the rotation of the Earth out of the realm of Paris and propelled it onto the world stage. Experiments using Foucault's pendulum began to be performed everywhere.

The very next experiment with Foucault's pendulum took place not much over a month after the first demonstration in the Panthéon, in May 1851, in the cathedral of the city of Reims, in France's fertile and prosperous region of Champagne, northeast of Paris.

This was, perhaps, a symbolic display of science, since the difference in latitude between Reims and Paris is not great, and therefore the period of rotation of the plane of the pendulum's oscillation could not differ by much, but there was certainly some difference.

Reims was interesting in other respects. Reims Cathedral was the place in which France's kings had been crowned, and so it had not only religious, but also patriotic meaning to the nation. Louis-Napoléon, who clearly liked public spectacles, pomp, and ceremony, could not turn down a display of his favorite physicist's pendulum in this regal and religious milieu.

Reims is an ancient city. As early as 250 A.D., pagans had been converted to Christianity in the Roman city, and a church had been erected here by the year 400. A century later, a charismatic monk by the name of Remi baptized France's king Clovis in the Reims church. In 1211, the magnificent cathedral, a masterpiece of Gothic art with over 2,500 statues, was built on the site of the ancient church. In this high cathedral, twenty-five French kings were crowned. Reims became a center for these coronations because the French wanted to continue linking their royalty with their Christian heritage. Thus each coronation reaffirmed the king's link with the line established by Clovis and Saint Remi at Reims.

A. M. Maumené conducted the experiment with the pendulum in the center of the Reims Cathedral on May 8, 1851. He used a steel piano string, 40 meters long, to suspend a pendulum bob weighing 19 kilograms. Maumené made the pendulum from lead that was poured into a glass shell with a steel rod placed inside it, to which the piano string was attached. Maumené immediately reported his results to Foucault in Paris, and the latter was overjoyed to find out that the result was in perfect accord with his sine law.

Others around the world quickly caught the Foucault pendulum fever. Within a few days, an experiment was carried out at another French cathedral, this time in the city of Rennes in Brittany, at almost the same latitude as Paris.

Next were English scientists, who carried out a pendulum experiment in the Radcliffe Library at Oxford. This experiment was extended over a period of three weeks. It was performed by Professor Morren, who used a wire 19 meters long and a bob weighing 30 kilograms. The results were not precise enough to determine whether they were in perfect accord with the sine law since there were so many visitors in the hall that the pendulum's swing was disturbed, introducing errors into the measurement process.

The experiment was repeated in Geneva with a 20-meter wire; then it was done in Dublin, Ireland; three more pendulum tests were carried out in the U.K.—in Bristol, in the York Cathedral, and in London. Across the Atlantic, Foucault's pendulum was demonstrated in New York that very same year.

A very important Foucault pendulum experiment was carried out in Rio de Janeiro in September and October 1851. Later described as "the marvelous demonstration in Rio," the experiment was the first one, other than the Panthéon experiment, to be reported by experts in the *Proceedings of the Academy of Sciences*.[56] The experiment was conducted by M. d'Oliveira. His pendulum was only 4 meters long, but it was kept going for two months without interruption. The line used to hang the bob of the pendulum was made of linen; its actual length was 4.365 meters, and it carried a spherical ball weighing 10.5 kilograms. While the report to the French Academy of Sciences did not include the time it took this pendulum to return to its starting point, there were many numerical results in the report. These, however, were not sufficient to make a decisive determination of exactly how well the findings accorded with the sine law, although it was clear that they did not contradict it.

Another experiment in low latitudes was done in Colombo, Ceylon. This one was carried out by Lamprey and Schaw. The pair used a line made of silk, which was a novelty. The line was 20 meters long. It held a 14-kilogram spherical bob made of lead, which was attached to the silk line with an iron ring. The silk held up well in the experiment. The results were good and accorded with Foucault's sine law.

The final test of Foucault's pendulum in 1851 took place—very significantly—in Rome. The experiment was, in fact, performed in the Jesuit Church of Saint Ignacius in the Vatican. Father Angelo Secchi (1818–1878) carefully implemented a 28.5-kilogram pendulum with a wire of 31.89 meters suspended from the high dome of the baroque church. This was an especially important experiment, since it was carried out in what was until then the bastion of anti-Copernican belief. Its successful completion

The demonstration of Foucault's pendulum in the Church of Saint Ignacius in Rome, 1851. (*Corbis-Bettmann*)

would signal a major change in the attitude of the Church toward science and particularly toward the theory that the Earth rotates.

The French had intended to keep the pendulum swinging in the Panthéon for many months. But, reversing the course, Louis-Napoléon issued a curious presidential order on December 1, 1851: The Panthéon experiment was to end immediately, and the Panthéon was to return to its function as a church.

11

———◆———

THE GYROSCOPE

Ignoring the overwhelming public reaction to his achievements and the cool reception he was receiving from the scientific community, Foucault was already working on the next proof he wanted to provide of the rotation of the Earth—one that would require much greater engineering skills than had the construction of the pendulum.

His greatest achievement that same year of the pendulum experiment in the Panthéon, 1851, was the invention of the gyroscope.

Foucault's biographer Stéphane Deligeorges traced this new invention designed to prove the rotation of the Earth to a discussion he believed took place between Foucault and the mathematician Louis Poinsot. The work of Poinsot was the most closely related to the problem Foucault was addressing. Poinsot had developed theories about rotating bodies, and Foucault is believed to have discussed with him his new idea, that of finding a new proof for the rotation of the Earth.

One problem Foucault saw was that many people had difficulties in understanding the complexity of the pendulum experiment: What does it mean that the "plane of oscillation" of the pendulum stays fixed while the Earth—the ground below the pendulum *and* the point through which the pendulum is suspended—rotates? In addition, there were difficulties in understanding the sine law: Why and how do the results depend on where the person observing the pendulum stands? In order to address the rotation of the Earth in a way that would be free of all these complications, Foucault sought another device. Through his discussions with Poinsot, Foucault came to realize that the oscillation of a pendulum was not the only type of motion that could demonstrate what he wanted. A rotational motion could be made to be independent of the rotation of the Earth. He concluded that a desirable instrument would be one that could provide an independent rotation of some object: a flywheel, or a disk, or a similar body. Foucault began to work on constructing such a device.

The fact that Foucault understood that a rotational device could serve to prove the rotation of the Earth is significant and shows, once again, that he was a careful scientist who understood the laws of nature in a fundamental and deep way. Foucault knew, without the use of equations or mathematical symbolism, that bodies in motion obey Newton's laws and will therefore maintain their inertial motion unless acted on by a force. This is what happens with the pendulum, which maintains its plane of oscillation regardless of the rotation of the Earth—if it is suspended freely without a torsion force on its string or wire making it turn with its support. Foucault knew that a similar phenomenon would take place with a rotating body, if it could somehow be suspended in space without any force moving it from its *plane of rotation*. What Foucault wanted, therefore, was what he called "a little isolated star"—a body whose motion could be viewed as independent of our Earth. According to Deligeorges, this notion had its origins in Foucault's discussions with Poinsot.

Once he knew exactly what he wanted, Foucault began to construct the desired apparatus. He turned again to his trusted friend Froment. The two of them worked together to first define the object Foucault needed to exhibit the rotational motion and—equally important—to devise the support system that would maintain the object rotating *freely* in space, a little isolated star, with no friction or torsion or any disturbing force.

What Foucault came up with was a little brass torus (a doughnut-shaped object) in whose center was a metal disk and through which was embedded a rod. This torus was held together in gimbals that allowed it to rotate freely and maintain its direction in all three dimensions of space. An attachment with toothed wheels allowed one to rotate the torus to give it high angular momentum through a series of gears and a handle. To this implement, which Foucault aptly named *gyroscope,* from the Greek, meaning "turn-see," he added a microscope that allowed one to see how the axis of rotation of the spinning torus turned with the rotation of the Earth. The microscope was needed because the implement used to give the gyroscope its initial spin could impart it a speed of about 150 revolutions per second; and through friction the energy slowly dissipated so that the gyroscope would only spin for about ten minutes at a time. During such a short interval of time, the axis of rotation would have only a small apparent deviation, and a microscope was useful for observing it.

Foucault enjoyed making the gyroscope. When he finished assembling the instrument, he invited Poinsot, who had inspired him to design the gyroscope and helped him with ideas about mechanics, to observe it. The two men spent two hours together, playing with Foucault's gyroscope: setting it in motion and observing the shift in position through the microscope. Recalling the event later in life, Foucault said: "I enjoyed myself very much that winter day."[57]

Here is how Foucault's gyroscope works. The spinning element maintains its direction in space. When a person runs the little machine with gears that make the torus spin, it is first spun in a particular direction. As the gyroscope continues to spin for some time, it maintains its direction in space in the same way that Foucault's pendulum keeps its plane of oscillation independently of the rotation of the Earth. This means that the axis of rotation of the torus, the direction of the metal rod about which it rotates, points in the same direction in space. To the observer, however, the axis of rotation of the torus seems to move in a circle as the Earth turns.

In this sense, the gyroscope is just like an amateur telescope with a little motor that makes the telescope rotate in a way that counters the rotation of the Earth, so that it can track whichever star it is aimed at (note that such telescopes are adjusted for latitude so that the arc they follow is the correct one).

When a person looks through the microscope attached to Foucault's gyroscope, it becomes evident that the axis of rotation of the torus is moving, and this apparent movement is a reflection of the fact that the Earth is rotating.

During the time Foucault was working hard on this new invention aimed at exhibiting the rotation of the Earth, a certain M. Lamarre from Brussels wrote a letter to the editor of the *Journal des Débats* poking fun at Foucault's alleged new project, calling the proposed gyroscope a "capricious little device." Foucault responded immediately, saying that he was waiting for the end of the vacation period so he could again present to the public a proof of the rotation of the Earth. He added: "But having learned that an honorable savant has found interest in that which engages me right now, I thought I should not waste a day before presenting to the public the facts acquired through my efforts on behalf of science."

Foucault's gyroscope indeed provided an alternative proof of the rotation of the Earth. Since the instrument was rather small—and required a microscope so that small momentary changes could be registered—it provided a less effective and less *dramatic* demonstration than did the pendulum.

An important application, however, was soon found for Foucault's new invention. When the gyroscope is aimed at a particular star, for example, it will track the apparent motion of this star across the sky. The reason is that the gyroscope maintains its direction in space as it rotates. Its axis of rotation does not change as its disk is spinning. A star sets when Earth's rotation makes it appear to drift across the sky toward the horizon and beyond. It is the rotation of the Earth that we observe when we see a star set. The same happens with the gyroscope: The gyroscope points in a fixed direction in space (pointing at a star), and as our planet rotates under it, the gyroscope keeps pointing in the same direction, but the direction seems to us to shift, since we are rotating away from it. Now, suppose that the gyroscope is initially aimed at the *pole star*. What would happen now?

We know that the spinning element of the gyroscope keeps the direction of its axis relative to the stars and is not affected by the rotation of the Earth—nor by any other motion of the system in which it is suspended.[58] The gimbals allow it to keep its direction as no force acts on it. So if the gyroscope is aimed at the North Star, it will continue to point at that star. This would turn the gyroscope into a compass. But the gyroscope could do more. With Foucault's new device it was possible to determine the direction of *true* north, which differs slightly from the position on the celestial sphere of the North Star. True north is the exact direction north, not the magnetic north, which differs from the true north because Earth's magnetic pole does not coincide with the geographic north pole. Above the true north pole is that point in the sky that does not appear to rotate but stays stable over time. Close to that point in the sky lies the North Star, but it does not coincide with it. The gyroscope could be aimed at that exact point, the celestial north pole. One could aim the axis of spin of the gyroscope at exactly that point in the sky, and this will turn the gyroscope into a very accurate compass—one that suffers neither from the deviation of the magnetic compass, nor from the deviation of Polaris (the North Star) from the celestial north pole.

The problem was only one of maintaining the spin of the gyroscope's element continuously. During the twentieth century, with the availability of electricity and electric motors, ships and airplanes all carried adapted versions of Foucault's gyroscope as their primary compasses.[59]

Foucault should have become very rich. The royalties that could have been paid him by all the companies using his invention should by all rights have made him a wealthy man. Unfortunately, the gyroscope remained ignored by industry for forty-seven years. In 1896, the Austrian engineer Ludwig Obry rediscovered Foucault's device and found a way of translating the gyroscope's directional properties into a set of instructions for steering a torpedo. The torpedo had been invented earlier in the nineteenth century, but it was inaccurate and often missed its target. Once Obry discovered how to direct it using a gyroscope, a company in Trieste bought the rights to it, and soon all the navies in the world were using gyroscope-directed torpedoes. Thereafter, the invention found its way into navigation. But all this was in the future, and in the meantime, the gyroscope's inventor, Léon Foucault, found himself unemployed.

As winter approached in that immensely fruitful year in Foucault's life, 1851, a year in which he stunned the world with his pendulum experiment and invented the gyroscope, an event was about to take place that would change life in France and would have worldwide implications.

12

———◆———

THE COUP D'ÉTAT AND
THE SECOND EMPIRE

Who was Louis-Napoléon Bonaparte? Émile Zola called him
"The Enigma, the Sphinx." And indeed, the man has remained
one of the most mysterious and least understood figures in history. His motives, desires, and goals are still ill understood a
century and a half after his time, and historians have found it
difficult to decipher his actions and plans. From a seemingly
aimless playboy, Louis-Napoléon was able to transform himself, after a series of false starts, into a position of leadership.
And now, he wanted to take one more step: toward absolute
power.

On December 2, 1851, an event took place in France that
would change history. It appears from newspaper articles of the
days leading up to December 2 that there was nothing unusual
to indicate that something very important was about to happen

in France. The National Assembly was embroiled in endless discussions of railroads. François Arago missed some meetings due to illness, and his son Emmanuel excused his absence to the Assembly. New fashions were paraded on the streets of Paris: Women wore long white dresses decorated with bright flowers or solid black or brown dresses with frills; men sported handsome jackets and bow ties; and hats were in style for both men and women.[60] Parisians, and French people everywhere, went about their merry daily lives. They suspected nothing.

But on December 2, Paris woke up to a shocking discovery. People found notices posted on every major building in the capital declaring a state of siege. Newspapers appearing later in the day, and those printed the morning after, all carried the same message on their front pages:

In the Name of the French People

The President of the Republic Decrees:

Article 1. The National Assembly is dissolved.

Article 2. Universal suffrage is reestablished; the law of
May 31 is abrogated.

Article 3. A state of siege is decreed.

> The Élysée Palace,
> 2 December 1851,
> Louis-Napoléon Bonaparte[61]

There followed an order to the population to stay calm and maintain the peace and to offer help to the authorities doing their work.

But the peace was not maintained. The French people did not easily give up their hard-won civil rights. Many of them gave fight. The president had prepared for this by arresting his opponents at dawn that morning and having them hauled off to jail. But enough citizens remained who were ready to resist the coup d'état.

Over the following days, newspapers reported about the fighting in the streets. The *Journal des Débats* said that "The students of the Latin Quarter have built barricades on rue Saint-André-des-Arts," and over the next few days barricades were reportedly put up on the other side of the Seine as well, in the working-class neighborhoods of the rue Saint-Martin and rue Saint-Denis. Every day for a week, the minister of war issued orders on the front pages of the newspapers to the citizens not to resist the soldiers and to help remove the barricades. By December 7, Paris began to quiet down, and within a few more days, other cities and towns around the country saw the resistance quelled as well.

It was now clear why Louis-Napoléon Bonaparte had closed the Panthéon the day before the coup d'état and decreed that it be reverted back to its function as a church. He was making an effort to appease the Church to prevent it from opposing his move toward absolute power.

On December 8, the president addressed the nation on the pages of the newspapers saying: "Society is saved. The first part of our task is over. Calm will bring a new era to the Republic." The last statement was especially strange, since the purpose of Louis-Napoléon Bonaparte's coup d'état was to do away with the

Republic altogether and to return France to the form of government of his late uncle's empire. It became clear that this indeed had been his plan all along. Louis-Napoléon did not work well within a democratic system. He had a stubborn, autocratic nature, and he wanted to rule alone.

"A police operation that was a little rough," is how one of the people responsible for carrying out Bonaparte's coup d'état described it. In reality, many people died or were wounded on the streets of Paris and elsewhere in France on or after December 2. Some historians later said that Louis-Napoléon suffered from guilt over the operation throughout his life.[62] And in many ways, this "rough police operation" was not in keeping with his character, although the true nature of the prince's character still remains a mystery.

Within days, the president called for a plebiscite, a vote by the people of France on a new constitution inspired by the one created by the French Revolution, in which the state would be headed by a president with limited powers and a term of ten years. Voting was to take place on December 20 and 21. The vote was in favor of Louis-Napoléon's program. This success would underscore his system of government. The prince despised the National Assembly but believed in a form of democracy in which the president is elected directly by the people and governs by the authority vested in him by the voters—rather than their representatives. He would continue, at least in principle, to seek the people's direct approval for his programs.

• • •

In a year, again on December 2—which was a significant day for Louis-Napoléon as the anniversary not only of the coup d'état but also of his illustrious uncle's victory at Austerlitz—Louis-Napoléon had the French people vote him Emperor of the French, thus establishing France's Second Empire. He took the name Napoléon III, to stress his point that he was the heir to his uncle Napoléon Bonaparte's dynasty, despite the fact that monarchies and a republic had intervened. This choice also implicitly acknowledged that Napoléon's son—had the empire been restored during his short lifetime—would have had the right to be called Napoléon II.

Louis-Napoléon now needed a wife. Only a married head of state could garner the respect of his people as well as foreign rulers and diplomats to the degree Napoléon III felt he should. Years as a perennial bachelor among international aristocrats and educated high society convinced him that marriage was essential to his success—despite his well-known promiscuity. And his chosen bride, he knew, should lack a similar past and should be high-born. Thus despite her ambition, her heavy investment in him, and her tears, the beautiful and loyal Miss Howard had to go. Louis-Napoléon was a decent man, despite what some would later say, and so he paid her off handsomely, and only after she had left France to return home to England did he begin in earnest his search for a bride.

After months of courting princesses and noble young ladies from around Europe, he settled for Eugénie de Montijo (1826–1920), Countess of Teba, a twenty-six-year-old Spanish noblewoman, and married her on January 29, 1853, in the Cathedral of Notre Dame in Paris. Eugénie had a strong personality and in his last years and after his death would exert considerable influence in France. But during most of the Second Empire, Louis-Napoléon would share his absolute rule with no one.

Soon afterwards, Louis-Napoléon moved from the Élysée Palace, the presidential residence he had been occupying, into his grand new apartments in the Louvre. This move, in many ways, symbolized the lavish, showy style that would characterize the Second Empire. Louis-Napoléon began a magnificent renovation of the Louvre Palace. He built new wings and joined the Louvre to the Tuileries Palace, which was located at the site of the present-day Tuileries Garden. The northwest wing of the Louvre became Louis-Napoléon's private apartments. Visitors to the Louvre today can still see them. They are wildly opulent, decorated in Louis XIV style—ornately carved, heavily gilded, lavishly furnished. There is a dramatic large hall with a high ceiling decorated with paintings of the emperor and his empress, Eugénie. Light sparkles from the large crystal chandelier, and the red velvet furniture is clearly made for a king. The dining room has a long, ornate mahogany table and is lined by two long rows of chairs, one on each side. The walls are decorated with scenes of the hunt. It is easy to imagine the emperor presiding over an elegant dinner party. The luxurious private quarters and bedrooms contain royal furniture and the original beds of the Bourbon kings.

The Louvre was not the only site of lavish building under the new empire. Louis-Napoléon built impressive new bridges on the Seine and palaces and large public buildings—in fact, he rebuilt the entire city of Paris as well as France as a whole.

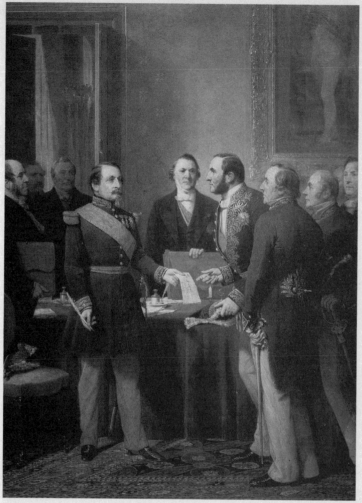

Napoléon III (left) meeting with Baron Haussmann, in a painting
by Adolphe Yvon. *(Corbis-Bettmann)*

In June 1853, Louis-Napoléon appointed Georges Haussmann Prefect of the Seine. Through Baron Haussmann's work
in this capacity, the empire embarked on a massive construction
operation. All the wide avenues and boulevards we see in Paris

today are the result of this major effort during the Second Empire to renovate—to remake, really—the city of Paris. The following passage gives an idea about just how much Paris today is the outcome of the transformation it underwent during the Second Empire under the direction of Baron Haussmann.

> Imagine Paris, the city of light, without clean air, pure water or proper drains. Take away its great highways, the boulevards Saint-Germain, Saint-Michel, Malesherbes, Magenta, Voltaire, Haussmann; the great crossing places of the Étoile, Republique, Trocadero, Opéra. Put back the slums of the Place du Carrousel into the gardens of the Louvre. Cover the Île de la Cité with a labyrinth of medieval houses, swarming up to the steps of Notre Dame. . . . Demolish many schools, the churches of St. Augustin, Trinité, and St. François-Xavier, and the theatres of Châtelet . . . entire residential *quartiers,* public buildings, museums, and part of the Louvre. Then you are left with some idea of Paris before Georges Haussmann became Prefect of the Seine.[63]

Haussmann turned an overgrown medieval town into a modern, thriving city. He built all the great boulevards we see in the city today. A large boulevard in the north part of the city was named after the man who made it: Boulevard Haussmann. To build it, the Prefect of the Seine destroyed, among many old buildings, also the house in which he had grown up. Parisians like to say that the wide boulevards that are so characteristic of the French capital were built so that barricades could never again block the streets, pointing out that it is easy to block a narrow street but difficult to barricade a wide boulevard. While this may be true, Bonaparte's other projects, all designed for the ostentatious glory of France, point to a wider appeal than the defense of the regime. It is worth noting that the first of the famous large department stores of Paris, *Au Bon Marché*, was inaugurated at the beginning of the Second Empire, in 1852, and the other *grands magazins* followed.

Louis-Napoléon had formed a plan for his empire long before the coup d'état. As early as 1839, he wrote:

The Napoléonic idea manifests itself in different ways in the various branches of human ingenuity: it is going to make agriculture flourish, it will invent new products, it will exchange new ideas with other countries. The influence of a great human genius, similar to the influence of divinity, is a fluid that spreads like electricity, elating the imagination, making hearts palpitate, as it touches the soul before it persuades the mind.[64]

Under Louis-Napoléon's leadership, France developed a modern banking system. The Second Empire saw an unparalleled expansion of the nation's railway system, which was one of the emperor's most passionate projects. Louis-Napoléon's vision of a progressive state also included reductions on customs dues and taxes. He wanted the French citizen to have more money and to enjoy a higher standard of living. People did well under his rule, especially during the golden decade from 1852 to 1862. They built luxury homes overlooking the elegant new boulevards, they dressed lavishly and enjoyed the theater, music, and strolls in the parks and gardens. The French economy thrived and expanded during the early years of Napoléon III's rule. There was a general feeling in the country of a perpetual pageant, a carnival, a parade.

The emperor had a clear vision of the new France he was creating. He wanted to build a modern state founded on progressive ideas. He believed in social values and democracy as manifested by universal suffrage. But he also wanted to rule alone. He never let any of his advisers, members of his cabinet, prominent citizens, leaders of industry, or even the empress (until he was weak and ill) influence his decision-making in any way. He listened to people, and he paid his aides and ministers very well, but ultimately there were no illusions about democracy or freedom of expression or meaningful participation in the running of the country. Louis-Napoléon's decisions were his own. And while he believed in the right of the citizens to vote, once the votes were counted, he alone was the ruler of France. Louis-Napoléon replaced the National Assembly he had abolished with the Legislative Corps, whose president and vice president were chosen by the emperor and whose membership was small and powers curtailed so that the body could not challenge the emperor's decisions. The discussions of the Legislative Corps were not reported in the newspapers. And the papers themselves were subject to censorship: There was no free press during the Second Empire. Similarly, the rights of citizens to congregate, to form organizations, or to act in any way that was counter to the government's objectives were curtailed and outlawed.

Externally, Napoléon III pursued the idea of France as an empire by seeking expansion and influence through war. Here again he was successful at first. His meddling in Italian affairs won him Nice and Savoy, although later he lost France the region of Alsace-Lorraine. He entered the Crimean war on the side of Britain and Sardinia and Turkey, against Russia. Having won, France increased its influence in Europe. Louis-Napoléon got France into Indochina, gained it a foothold in Lebanon, and had a debacle in Mexico. He had established an Austrian royal, Archduke Maximilian, as Emperor of Mexico, only to find him executed by rebels in 1867.

Louis-Napoléon believed in education, something that had been denied him. During the Second Empire, the number of students in colleges in France tripled from about 50,000 to over 150,000 per year. Great advances were being made in the areas of science and technology throughout the world during this period. In England, the invention of the steam engine was promoting the development and manufacture of ingenious machines of all sorts. Astronomy and physics were expanding our knowledge of the universe at a rate that would only be matched during the end of the twentieth century. John Frederick Herschel, the preeminent British astronomer, was making great discoveries of stars, and Michael Faraday was developing the field of electricity and magnetism. Napoléon III wanted his France to become the world leader in science and technology.

Having taught himself science at the Ham prison, Louis-Napoléon likely identified with the other self-taught scientist, Léon Foucault. Both men shared a past of great struggle against a hostile elite. Each of them had to fight for what he wanted to achieve and had a supreme confidence in his abilities. Perhaps because of this similarity between the two, their shared love of science, and the natural affinity between them, Napoléon III was willing to do for Foucault something he would not have done for any other scientist: to invent a job for him.

13

❖

AN UNEMPLOYED GENIUS

Léon Foucault was a product of this great new age of science and technology. He was naturally gifted in the areas required for work in science and engineering, and he knew it. Foucault had already participated very actively in a number of fields of study and discovery in this exciting time in history. He had worked on photography, improving the daguerreotypes; he measured the speed of light; he invented new lighting systems; and he provided incontrovertible proof of the rotation of the Earth. And he created the gyroscope—an invention on par with the greatest technological advances of this age of technology and progress. Now he was eager to continue his work. But he was suddenly unemployed. With the exception of writing for the *Journal des Débats,* Foucault had nothing to do with his time—he lacked an

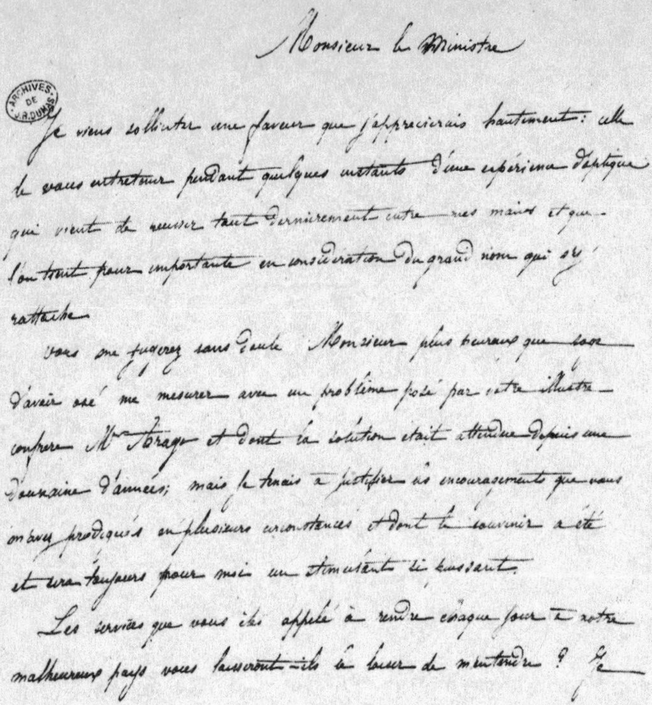

[Handwritten letter in French, partially legible]

Monsieur le Ministre

Je vous sollicite une faveur que j'apprécierais hautement : celle de vous entretenir pendant quelques instants d'une expérience d'optique qui vient de réussir tout dernièrement entre mes mains et que l'on tient pour importante en considération du grand nom qui s'y rattache

vous me jugerez sans doute Monsieur plus heureux que sage d'avoir osé me mesurer avec un problème posé par cet illustre confrère Mr Arago et dont la solution était attendue depuis une douzaine d'années; mais je tenais à justifier les encouragements que vous m'avez prodigués en plusieurs circonstances et dont le souvenir a été et sera toujours pour moi un stimulant si puissant

Les services que vous êtes appelé à rendre chaque jour à notre malheureux pays vous laisseront-ils un loisir de m'entendre ? Je

outlet for his tremendous productive energy and his unparalleled abilities.

Foucault was not a quiet man patiently waiting in the wings to be recognized. There are many letters in his biographical file at the French Academy of Sciences that attest to his being an indefatigable applicant. As early as the 1840s, when he was a laboratory assistant doing research on microscopy and light regulators, he was writing letters to the Minister of Public Instruction asking for support for his projects. On May 31, 1850, Foucault wrote to the minister invoking his work on behalf of Arago, hoping that this would get him financial support. It is not

One of Foucault's letters to a government minister. (*Académie des Sciences, Paris*)

clear that these letters helped him in any way. Still, he kept looking and applying and requesting an appropriate position. Finally, in 1855, he was successful. He had someone who could help him: Louis-Napoléon Bonaparte.

Following the establishment of the Second Empire on December 2, 1852, Louis-Napoléon required prominent leaders to pledge allegiance to the empire. In the Observatory and the Institut de France and the Academy of Sciences, François Arago refused to swear allegiance. He retreated into the Observatory and conferred with his close friends and associates. Arago expected that, as a result of this move, he would lose his Directorship of the Observatory.

• • •

Urbain Le Verrier (1811–1877), Arago's junior by twenty-five years, was undoubtedly France's most celebrated astronomer of all time. Le Verrier was born on March 11, 1811, in St.-Lô, Normandy, to a family employed in local government. Le Verrier studied at the École Polytechnique in Paris, where he enrolled as a student of chemistry in 1831. He later changed his field of study to astronomy and in 1837 was appointed junior astronomer at the École Polytechnique. He was interested in extending the work of the great Laplace in celestial mechanics, the field that addresses problems of mechanics in astronomy. In particular, he was fascinated by the issue of the stability of the solar system: Why is it that all the planets move around the Sun in an ordered and regular fashion without a planet suddenly veering off into space? Le Verrier made some progress in his study of the problem of the stability of our solar system, and his work came to the attention of François Arago at the Observatory.

Arago thought that the brilliant young astronomer might help him find clues to what was then one of the biggest mysteries in astronomy: Why was the planet Uranus deviating from what seemed to be its normal planetary orbit? What was disturbing this planet from its regular course through the night sky?

In 1781, William Herschel discovered the planet Uranus. Observations of this planet allowed astronomers to determine the elliptical orbit of Uranus and produce tables of future positions it should take. But astronomers soon discovered that Uranus deviated from these predicted positions in the sky. Astronomers put forward several guesses as to the reason for this mysterious behavior by the planet. In deriving its law of motion they used Newton's inverse-square law, which said that the gravitational attraction between two bodies depends on the masses and the distance between them: It increases with mass and decreases in proportion to the square of the distance. Perhaps this law did not hold at distances as great as that of Uranus from the Sun? Another theory was that a comet had hit Uranus just after it was discovered, making it deviate from its expected course. A third theory, one that Arago favored, was that Uranus was being pulled away from its normal orbit by another, yet-undiscovered planet, and therefore, astronomers could not correctly predict where it should be: They were missing in their calculations this additional mass belonging to the unknown planet.

Urbain Le Verrier had studied methods of mathematical astronomy and had become an expert in this area. Mathematical astronomy is the field in which scientists carry out theoretical calculations using pencil and paper. They manipulate equations, such as Newton's laws of motion, and make predictions based on their theories. Then these computations are put to the test by comparison with observations of the heavens produced by observational astronomers. In 1845, under Arago's supervision at the Paris Observatory, Le Verrier produced a theoretical paper on the movement of the planet Uranus. In November of that year he presented his paper to the French Academy of Sciences, and the paper came to the attention of the British astronomer George Biddell Airy (1801–1892). Le Verrier continued his work and in March 1846 was able to predict where the new planet should be in the sky. He computed that the missing planet, if indeed there was one, had to be within a few degrees of celestial longitude 325 degrees as seen from the Sun.

While Le Verrier was unaware of it, an English astronomer, John Couch Adams (1819–1892), had been working on the same problem for several years. He finally reached a solution to the problem in late 1845. He determined that the missing planet should be found at about longitude 323 degrees and 34 minutes from the Sun. Adams contacted Airy, but the two did not meet because of a sequence of misunderstandings. However, Adams managed to leave his results for Airy. But Airy only paid attention to Adams's paper once he received Le Verrier's paper some time later. This finally stirred him to act, and he contacted astronomers in Britain asking them to look for the missing planet. A low priority was given to the search, and this ultimately cost Adams his recognition as the first to predict the position of the missing planet. For while the British were wasting time, Le Verrier persuaded astronomers at the Berlin observatory to look for the missing planet. In September 1846, astronomers in Berlin found a "star" that was not on their star maps at a position close to where Le Verrier told them to search for the new planet. They had just discovered the eighth planet of the solar system, which Victor Hugo suggested be named Le Verrier, but which we know today as Neptune. The English never accepted Le Verrier's discovery—wanting, after having bungled the opportunity for primacy, to give sole credit to Adams. But today most recognize Le Verrier as the discoverer of Neptune. Overnight, Le Verrier became a hero in his native France, and his career was launched.

• • •

Unlike Arago, who was also a product of the École Polytechnique, Le Verrier did not harbor any anti-Empire feelings. Having been honored by France for his immensely important discovery, Le Verrier was wined and dined by royalty, he was awarded the Legion of Honor commission from King Louis Philippe, was appointed to the prestigious Board of Longitudes, and was given the important position of chief of astronomical research at the Paris Observatory. Everyone expected that with Arago's refusal to pledge allegiance to Napoléon III, Le Verrier would be named to replace him. But despite the unparalleled affront to his rule, the emperor did not dismiss the old and ailing Director of Observations at the newly named Imperial Observatory. Arago was left to continue his scientific work both at the Observatory and at the Academy of Sciences, although he repeatedly said he wanted to step down from his position as Perpetual Secretary of the Academy because he was feeling increasingly weak. On August 22, 1853, Arago came to the Academy for the last time. Despite his bad health, he left from there to go to the Eastern Pyrenees. A dear family friend had died there, and there was a wedding of a relative and other family business that required his presence. The voyage to his home district was arduous: He had to make several train connections, as well as travel by carriage. He arrived at his family home exhausted. But soon he recovered and spent the rest of his time chastising his nephews for not having published their works, prodding them to do so. He was back to his old authoritarian ways. But physically, Arago was getting weaker. Rumors of his death began to circulate in Paris early that fall. But Arago struggled to stay alive for another month and died at dawn on October 2, 1853. When the Academy of Sciences met the next day, its members all rose and stood for a moment of silence to honor their deceased Perpetual Secretary. His eulogy in the *Proceedings* included these words:

> His name, illustrious for fecund and original discoveries, shall be inscribed in the splendor of science among the greatest, among Herschel, Young, Watt, Fresnel, Ampère, and Gay-Lussac.[65]

The emperor, somewhat relieved that he never had to dismiss the very popular scientist and statesman, now made Le Verrier the new Director of Observations—as had been his plan for some time. Thus began a new era for science and a new phase of the emperor's relationship with science and with scientists. Urbain Le Verrier and Napoléon III would become close friends.

When it came to the attention of the emperor that his favorite scientist Léon Foucault was out of a job, he decided to do for him something that he would not do for many other people: Louis-Napoléon would *create* a job for him, the position of Physicist Attached to the Imperial Observatory in Paris. Foucault was made to understand through the appropriate official channels that he should write a letter to the new Director of Observations,

Urbain Le Verrier, and describe to him what he saw as the role of such a Physicist Attached to the Observatory. Then, the French administration would take care of his application for this position.

Foucault immediately set to work. On August 19, 1854, while staying at Maison Briffard in Dieppe, on the coast of Normandy, he wrote Le Verrier a long letter in which he detailed how he saw the role of a Physicist Attached to the Observatory. He said:

Monsieur,

I am going to try to indulge your counsel of defining in a few words the services that I could provide to science as Physicist attached to the Observatoire de Paris. In my view, such services could fall into two distinct categories:

First, at the Observatory a physicist could naturally put all the resources of modern physics at the service of astronomical observations . . .

And second, to attack certain problems of physics whose study requires astronomical observations . . .

In the first category: apparatus, applications of telegraphy, production of images, etc.[66]

The letter continued over four pages, giving detailed descriptions of what Foucault thought he could do.

But this letter, clearly the result of great thought and insight, was apparently not enough. There is no evidence that Le Verrier responded positively, and so Foucault enlisted others to write to

Le Verrier on his behalf. One of them was his old professor, Alfred Donné. Another was Foucault's close friend Jules Regnauld (1820–1895), with whom he had conducted experiments on the perception of color some years earlier.

When all seemed not to come to fruition, Foucault wrote Le Verrier: "I have no ambition of occupying an eminent post. All I want is to preserve my facility to work and to accomplish modestly my destiny."[67]

Some time later, the order came down directly from the emperor, and Le Verrier appointed Foucault as Physicist Attached to the Imperial Observatory in Paris. The emperor would not regret this move. But before turning his attention to astronomical research, Foucault had another opportunity to showcase his marvelous experiment demonstrating the rotation of the Earth.

In 1855, France hosted the first Universal Exhibition in Paris. At 1 P.M. on May 15, 1855, the Emperor Napoléon III and the Empress Eugénie stood on the 250-meter-long path at the Palace of Industry in Paris. In front of them were representatives of thirty-four nations, who came as invited guests of the empire to the Universal Exhibition, the first ever to be so named. The emperor exchanged a few words with the guests nearest to him, and an orchestra of 150 musicians played a tune composed by Louis-Napoléon's mother, Queen Hortense.

For the occasion, Léon Foucault had prepared a magnificent

pendulum. It hung from the ceiling of a high hall and had a bob made of iron. For this new pendulum, Foucault added a novel invention to allow the guests who came to this event to see the pendulum swing continuously with no interruptions. Until then, when the pendulum started, it would continue swinging until the energy dissipated after some time. His new pendulum was powered by an electromagnet, so that the energy never dissipated and the pendulum kept swinging until the power to the electromagnet was turned off. The new device was inspired in its inception by Foucault's previous work on electrical regulators. But here, he had a problem. When force was applied by the electromagnet to the pendulum, it would swing out of control. The trick Foucault discovered in his quest for a pendulum with perpetual motion was to apply the force from the electromagnet only while the pendulum's bob was heading down. Once the pendulum reached the vertical, the electromagnetic force was turned off. This ensured a regular and continuous motion without chaotic buildup of amplitude. Another problem the inventor had to solve was one of making the electromagnet follow the direction of the pendulum's swing without itself affecting it. Foucault solved this problem, too, and the results were impressive.

14

───◆───

THE OBSERVATORY PHYSICIST

Louis-Napoléon Bonaparte was proud of his scientist and his achievements. The emperor had created the position of Physicist Attached to the Imperial Observatory, and he continued to follow Foucault's career with interest, occasionally inquiring about his progress and marveling at his inventions and discoveries. In 1855, a letter from the emperor's Cabinet informed Foucault that Napoléon III was "deeply interested" in his work and would support financially all his future research.[68]

Through this relationship between emperor and scientist, another friendship developed. This was the friendship between Napoléon III and Foucault's superior, Urbain Le Verrier. Le Verrier was a person of high standing—he was not only the Director of Observations and Perpetual Secretary of the Academy of Sciences, but also France's scientific hero—and thus the emperor could enjoy a close friendship with him. Napoléon also knew that Le Verrier, unlike the late Arago, was loyal to him and that he didn't concern himself with questions of democracy or the powers of the emperor.

Louis-Napoléon continued to be interested in science. He read prodigiously about scientific issues and was especially interested in problems of astronomy. Le Verrier would send the emperor books and papers on recent scientific discoveries, and Napoléon would respond, writing him letters in which he discussed those issues that were near to his heart. The emperor was particularly interested in the question of time, in determining exact time using astronomical observations. He wanted to know how to find the date and time of the equinoxes, and he was interested in astronomical tables and star catalogs. A personal correspondence developed between the two men. Napoléon wrote Le Verrier warm letters, asking him questions about these issues, telling him about scientific papers he had read, and discussing ideas. He would end his letters by declaring his sincere feelings of personal friendship to the astronomer. And he would sign simply as Napoléon. One such letter is shown on the next page.

Louis-Napoléon's interest in astronomy extended beyond his own empire. Le Verrier carried out a correspondence with many astronomers around the world, and John Herschel wrote him a letter (of an uncertain date) from England, in which he said: "In reply to your confidential enquiry, my work, as I have every reason to presume, is by this time actually in the hands of your illustrious Sovereign, having been presented to him by the Duke of Northumberland in the joint names of his Grace and myself."[69]

A letter from Napoléon III to Le Verrier. (*Institut de France, Paris*)

Napoléon's deep interest in science not only established the career of Léon Foucault against the objections of the mathematical community of France, but it also provided an impetus for scientific work both in France and abroad. The emperor's positive outlook was a guiding light for science and technology during his reign. It also helped provide financial support for scientific investigations. Thanks to the emperor's interest, science flourished in France during the Second Empire and led to technology that improved the economy and the daily lives of the French people.

• • •

The fantastic success of Urbain Le Verrier signaled a new age in astronomy and in the physical sciences in general. Le Verrier was able to discover a new planet not by tireless nights at the telescope, searching every patch of the sky—as, in fact, is done by most comet-hunters today—but rather through mathematics. Le Verrier's achievement is a crowning glory for French theoretical analysis. Here was a bright young mathematician, working with formulas and mathematical derivations through which the location of a missing planet was predicted with such a high degree of accuracy that instructions to observational astronomers could be given, telling them exactly where to look for the planet, and so leading to a discovery. Le Verrier brought astronomy to a new age of mathematical analysis.

This success further reinforced the camp of the physicists and mathematicians, at the Academy of Sciences and elsewhere, who believed that scientific study should be based on solid mathematical analysis. And as a consequence of this rise of theory in science, Foucault's own way of conducting scientific investigations, using an empirical approach, would become even less favored. And yet Foucault would continue to amaze the world of science with his discoveries and inventions. Finally in a position to continue his work, he began to produce prodigiously.

As the new Director of Observations, Le Verrier recognized that the Paris Observatory needed new telescopes if it were to compete effectively with other observatories around the world. The age saw the proliferation of observatories and telescopes as humanity turned its eyes to the skies. There was so much to be discovered: planets, comets, stars. Louis XIV's dream of opening up the heavens to human observation and knowledge was be-coming a reality during this heyday of astronomy and science. The Imperial Observatory had to keep up if it were to continue its legacy as the place where the greatest discoveries had been made during the early days of astronomy. Le Verrier wanted his new Physicist Attached to the Observatory to build a 74-inch telescope.

Foucault set to work on the new telescope and invested in this new project all his energies, which had been pent-up over a period of forced idleness that followed his last great achievement, the invention of the gyroscope. Within a short time, Foucault came to the conclusion that a *reflecting* telescope was much better than a *refracting* one. A reflecting telescope is one that uses a curved mirror to focus light, while a refracting telescope uses a large lens to do the same thing. The idea for the reflecting tele-scope was proposed in 1663 by the Scottish mathematician James Gregory (1638–1675). The design used two concave mirrors to concentrate and reflect light into an eyepiece. Gregory's idea was to build a telescope that would not have to be very long and heavy, as refracting telescopes tended to be. In 1672, in France, an astronomer named Laurent Cassegrain (1629–1693), of Chartres, made an improved reflecting telescope, which used a convex secondary mirror. In 1668, Isaac Newton in Cambridge built a reflecting telescope of a different design. Newton's tele-scope used a flat secondary mirror to divert light at 45 degrees to an eyepiece. Newton built several similar telescopes, some of which he presented to the Royal Society. Both designs, the New-tonian design and the Cassegrain design, survive to this day.[70]

As judged by the success of the reflecting telescopes today, Foucault's choice of a reflecting telescope was a very good one. His work was at the leading edge in telescope design. Foucault made several telescopes, some of which can still be seen at the Paris Observatory. Foucault's largest telescope was an 80-inch reflector, and it is still used at the Observatory of Marseille. The south of France enjoys clear skies most of the year, in contrast with the often-cloudy weather in Paris. Thus as astronomy expanded and developed as a science during the nineteenth century, it became increasingly more difficult to plan observation sessions in Paris because of the weather, and the center of astronomical activity moved south. Today, the Paris Observatory is still a very impressive landmark with its old buildings and facilities, but actual observation of the skies takes place at high mountain peaks around the world. Marseille uses the Foucault telescope for observations, although this observatory's usefulness has also been surpassed by the higher observatories with their much larger telescopes—all of them using the reflecting, rather than refracting, designs.

Foucault's great discovery in the area of astronomical instrumentation was a method of silvering the mirrors for reflecting telescopes. Reflecting telescopes that had been made up to that time used bulky, heavy mirrors. This limited the potential size of these telescopes because the weight of the mirror could collapse the telescope. Foucault inaugurated a new method of applying a layer of silver directly to the front of the telescope's mirror, rather than a mercury amalgam that was typically applied to the back of the mirror. Foucault's telescopes built this way were lighter and of better light-gathering quality than earlier telescopes, as evidenced by their use today, a century and a half after his time.

Foucault also combined his interest in machines with his interest in telescopes. He designed a motor that would allow the telescope to counter the effects of the rotation of the Earth on astronomical observations. The motor continuously turned the telescope so that it could remain aimed at a given star despite the apparent motion of the celestial sphere. Most machines available at that time did not provide constant motion but tended to accelerate. Foucault developed a regulator for a small motor, which controlled the rate at which the motor rotated the telescope, keeping this motion constant and precisely matched to the apparent movement of the night sky.

As Physicist Attached to the Observatory, Foucault made many discoveries. The Paris Observatory still has a building designated "Foucault's Laboratory." Here, the indefatigable scientist invented a mercury switch for wind induction, designed a mercury current interrupter, and built a photometer. Foucault also demonstrated the conversion of mechanical work into heat energy by turning with a crank a copper disk placed between the poles of an electromagnet and measuring the amount of heat produced in this way. Foucault discovered the existence of eddy currents in a copper disk moving in a strong magnetic field. Such currents are called *Foucault currents*. Foucault also developed a series of mechanical regulators, which he hoped to sell to industry, but the market did not show much interest in them. His most important work at the observatory, however, was in developing mirrors and prisms and other components used in telescopes.

• • •

Le Verrier did not stop his scientific work after assuming administrative duties both at the Observatory and the Academy of Sciences. France's great astronomer, who could deduce the existence of a missing planet using strictly theoretical calculations, now assumed an even more ambitious project. Le Verrier embarked on a mission of estimating the masses of the planets in our solar system. He considered the average distance to the Sun a fundamental parameter of prime importance in astronomy since its value affected the estimates of planetary masses through theoretical computations based on this distance. This constant depended on another—one whose importance Le Verrier, living in the 1800s, could not fully appreciate: the speed of light.

Le Verrier therefore asked the Physicist Attached to the Imperial Observatory to redo his old experiment on the speed of light in a way that would give him the numbers he needed for his calculations. Foucault rebuilt the equipment he had used years earlier in his study of the speed of light and improved his design in a way that could yield results on the actual speed of light, as Le Verrier wanted. For this purpose, the Director of Observations gave Foucault the use of Meridian Hall, the site of his first public success with the pendulum. And again Foucault's friend and long-time associate Gustave Froment assisted him with the experiments. In September 1862, Foucault was able to report to the Academy of Sciences his results on the speed of light. His estimate was 298,000 kilometers per second, a number that is accurate to within 0.6 of one percent of the presently accepted estimate of this constant. The adopted value today for the speed of light is 299,792.458 kilometers per second. This value, without the decimals, has been known since 1947, and the accuracy to third decimal place has been obtained using lasers in later tests in the 1970s and 1980s. Foucault's estimate of the speed of light would only be improved twenty years after his experiment by the American physicist Albert Michelson (1852–1931), using the same method of a rotating mirror. Foucault's result on the speed of light allowed Le Verrier to estimate the distance to the Sun and, from it, to produce good estimates of the masses of the planets in our solar system.

Foucault's life at the observatory also provided opportunity for adventure. In 1860, Foucault and Le Verrier embarked on an expedition to Spain to observe the total solar eclipse of July 18, 1860. Foucault's task during this expedition was to photograph the sky during the eclipse. He had thus returned to his first scientific love, the science of photography. Foucault was very excited on this voyage. Being a careful scientist and one who tended to worry about every detail, Foucault kept urging Le Verrier and their assistants to move the site of their observation post from one place to another since he feared that at their original location, close to a mountain, they might be vulnerable to cloudy weather—the scourge of eclipse-chasers throughout history. His choice of a new location proved adequate and the team was able to observe the eclipse unhindered by clouds. The sight of a total solar eclipse moved Foucault and he described it in writing as follows:

The moon, gliding invisibly in space,
Advanced in slow and uniform steps.
She encroached on the solar disc,
Giving birth to a brilliant necklace
Growing larger by the minute.[71]

Foucault had come full circle with his work. Having done so much for science, he was now back working with photography in the romantic setting of a Spanish eclipse. The euphoria of watching one of nature's most awesome events, photographing it, and studying it scientifically made him contemplate nature's deepest secrets. Was the ether real? he asked himself. Did his pendulum swing honoring absolute space, staying faithful to the unseen ether?

While today we know that the ether (a hypothetical medium that permeates space) does not exist, the deeper answer to Foucault's question still remains a mystery. What is the secret of Foucault's pendulum? Why does it swing the way it does? What guides the pendulum in the continuous veering of its plane of oscillation?

The explanations of Foucault's pendulum have remained naïve and incomplete for decades. To shed some light on the mysteries of the pendulum, the world had to await the works of two great thinkers. The first was Ernst Mach (1838–1916), an Austrian philosopher and physicist who was only thirteen years old when Foucault carried out his experiment in the Panthéon but whose future work would make us view the pendulum experiment in a different, new way. The second was Albert Einstein (1879–1955).

To explain the result of Foucault's pendulum experiment (and other phenomena), Mach put forward a new concept, one that Einstein would later call *Mach's principle*. According to Mach's principle, Foucault's pendulum maintains its inertia—its undisturbed regular motion in space—with respect to a reference frame set by *the distant stars*.

Foucault's pendulum swings along its plane of oscillation, and that plane of oscillation remains fixed, not relative to Newton's "absolute space," but rather with reference to the entire conglomeration of all the masses in the universe. Somehow, all the distant stars and galaxies in the universe define a frame of reference within which the pendulum's plane of oscillation remains constant. According to Mach, if these other masses—the distant stars—did not exist, there would be no rotation of the Earth, and Foucault's pendulum would not appear to deviate from its original plane of oscillation. If Mach's principle is correct, then rotation cannot exist without an outside reference frame set by the aggregate of "all the masses in the universe" (with the importance of each mass decreasing with its distance). This is a deep, cosmic principle, which implies that the simple pendulum somehow "senses" the presence of all the other masses in the universe.

Albert Einstein admired Mach and his ideas. This is why it was Einstein himself who coined the phrase "Mach's principle." However, as Einstein developed his general theory of relativity, which he completed in late 1915, his ideas deviated from those of Mach. Einstein's general relativity is in a way placed between Mach's ideas and those of Newton.

According to general relativity, inertial frames of reference are determined by the local gravitational field, which is induced by all the matter in the universe. However, once a body is in an inertial frame of reference, the laws of motion governing it are completely unaffected by the presence of nearby masses. For example, the mass of the Sun determines the motion of the Earth, which is in "free-fall into the Sun" (the Earth's rotation around the Sun). But once we fix our coordinate system to the Earth itself, we cannot detect the gravitational field of the Sun. This fact has been demonstrated experimentally.[72]

General relativity thus says the following: In the absence of nearby matter, the inertial reference frame is determined by all the masses in the universe. When a large mass such as the Sun is brought close by, this mass changes the inertial frames so that they accelerate toward it. But the laws of motion in these free-falling frames show no effect of the surrounding mass distribution. General relativity and Mach's principle are opposite to one another in this respect. Which one is correct? This remains an open question, even though general relativity has been confirmed in a wide variety of situations.

The belief that the average motion of the rest of the universe affects the behavior of any single body presents many questions. Would the charge of the electron, for example, change if the number of particles in the universe were different? There are many such questions, and they remain unanswered. The relationship between what happens here and what happens in the entire cosmos is not understood. We don't know exactly what reference frame the plane of oscillation of Foucault's pendulum respects; and the true, philosophical nature of Foucault's pendulum remains a mystery.

Foucault came back from his marvelous voyage invigorated. He told his friends that there was so much in nature he still wanted to explore, so many wonders he wanted to study, and that he felt he had twenty years ahead of him in which to pursue the excitement of scientific discovery. But the eclipse expedition to Spain would be his last scientific voyage.

15

FINAL GLORY

In the years leading to his death, Léon Foucault finally received many of the honors he deserved. He was awarded a Doctor of Physical Science degree for his experiment comparing the speed of light in air and in water. In 1862, Napoléon III made Foucault an Officer of the French Legion of Honor. That same year, Foucault was elected for membership by the Bureau of Longitudes, a group that included the world's greatest scientists. And in London, the Royal Society elected Foucault as a foreign member. He was given similar recognition in Germany.

But Foucault was still not a member of the Academy of Sciences. That final honor—the ultimate sign of acceptance by the people whose approval and respect he desperately wanted—was finally granted him only three years before his death: in 1865. This honor came to Foucault at the end of a long and arduous process.

• • •

It began in 1851, right after Foucault's successful experiments with the pendulum. There is an interesting item in the *Proceedings* of the Academy of Sciences in 1851 dealing with Foucault, a description of the results of an election for membership in the Academy, appearing in the March 17 issue. Foucault's biographers have said that Foucault was rejected in his attempts to gain membership in the Academy until close to the time of his death. This is somewhat misleading. While it is true that Foucault was privately disliked by the mathematicians who were put to shame by his achievements, Foucault did have friends and admirers. We know this from the many obituaries written about him after his death by close friends and members of the Academy: Lissajous, Morin, Gariel, Gilbert, and Bertrand. These friends recognized his genius and supported his efforts to gain recognition.

The election for membership in the Academy took place in March 1851 to replace the chair of the general physics section vacated by the famous physicist and chemist L.-J. Gay-Lussac. In the first round of voting, there were seven candidates. Foucault's friends managed to get him on the secret ballot, and out of fifty-two total votes, he received eleven. He was tied in second place with Edmund Becquerel; and ahead of them, at the number one spot, was Charles Cagniard-Latour with twelve votes. Interestingly, Fizeau was tied with someone else at number four with eight votes in his favor. The fact that Fizeau was also nominated may have meant that the Academy was recognizing—in its own way—Foucault and Fizeau's work on the speed of light. Since no one got an "absolute majority," as required by the constitution of the Academy of Sciences, the process was repeated; it had to be repeated four times in total. On the fourth round there were only two candidates: Cagniard-Latour with thirty-four votes, and Léon Foucault with nineteen. This result constituted the absolute majority needed, and Cagniard-Latour was selected.

After being rejected for admittance into the Academy of Sciences, Foucault became even more determined, and he made it a point to keep applying again and again. Whenever a member of the Academy died and a position opened up, Foucault would immediately write a letter to the president of the Academy, asking to be voted on again for membership. His drive and determination to succeed were similar, in many ways, to those of Louis-Napoléon, and they remind us of the latter's efforts twenty years earlier to gain control over France despite the odds against him in Strassbourg, Boulogne, and in elections.

On February 22, 1858, Foucault wrote a letter to the president of the Academy asking to replace the deceased Cauchy. He wrote:

Mr. President,

The very regrettable death of the illustrious geometer M. Cauchy leaves the section of Mechanics with a vacancy that the Academy would like to fill. Many candidates have already made clear their intentions to apply for this position. The discovery of phenomena of relative movement that are produced in the presence of the rotation of the Earth constitute the titles that I would like to present in this circumstance.

In consequence, Mr. President, I express to you my desire that
you inform the Section of my request to consider me as a
candidate for the place just vacated within it.

I have the honor of being, Mr. President, with profound
respect your very humble and obedient servant.

Léon Foucault[73]

Foucault's application was rejected. After a further rejection
of an application for membership submitted five years later, Foucault wrote the following angry letter.

Paris, May 11, 1863

Mr. President,
A place was made vacant in the Physics section by the death
of Mr. Despretz. Having been admitted to the list of presenters, I
considered myself qualified to be a candidate. I had the confidence
that in these particular circumstances, the Academy would
pronounce to adjourn [meaning stop the process of reading more
applications]. It had decided otherwise.

I dare request of you, Mr. President, to be my conduit to the
Academy and to express to it all my regrets for the fact that it has
decided not to offer me the recognition for my modest research.[74]

According to Foucault's biographer Stéphane Deligeorges,
Foucault continued on a frenzy of scientific activity until he
found out, in 1857, that—once again—he had been rejected for
membership in the French Academy of Sciences.[75] While his first

few rejections spurred him on to increase the level of his scientific work, the rejection of 1857 was more devastating and stopped much of his activity. It is unfortunate that these events took place. How could the French scientific community treat Foucault this way? How could they completely ignore the work of the man who—after centuries of ignorance about the state of the universe—finally brought us such a clear, simple, and elegant proof of the rotation of the Earth, accomplishing a goal other scientists had said was impossible? But this was the shocking state of French science in those days. Someone who was not in the inner circle of science in the French capital did not have a good chance of entering the club, regardless of achievement. In the years until the devastating rejection of 1857, however, Foucault accomplished a great deal.

Foucault continued to apply for membership in the Academy of Sciences. On January 9, 1865, Foucault wrote his by now usual letter. "The place having become vacant by the death of M. Clapeyron, it seems to me, should become filled by a person in mechanics . . ." He went on to concede that others would argue that the place should be filled by a "geometer." This seems to have been Foucault's weakest application—he knew they were looking for a mathematician. And the letter seems almost to have been written perfunctorily by someone who believed he had no chance.

Surprisingly, it worked. The Academy approved his membership, and on January 23, 1865, Napoléon III issued his imperial decree making Foucault a member of the Academy:

Ministère
de l'Instruction publique
et des Cultes

Secrétariat général

2ᵉ Bureau.

5602

Napoléon, par la grâce de Dieu et la volonté nationale, Empereur des Français,

À tous présents et à venir, Salut.

Sur le Rapport de notre Ministre Secrétaire d'État au département de l'Instruction publique et des Cultes,

Vu l'extrait du procès verbal de la séance tenue le 23 Janvier 1865 par l'académie des Sciences de l'Institut Impérial de France,

Avons décrété et décrétons ce qui suit:

art. 1ᵉʳ

L'élection que l'Académie des Sciences de l'Institut Impérial de France a faite de M. Foucault, pour remplir la place d'Académicien devenue vacante dans la section de mécanique, par suite du décès de M. Clapeyron, est approuvée.

Art.

Napoléon, by the grace of God and the national will, Emperor of the French,

To all present and to come, Salutations.

On the report of our Minister Secretary of State at the Department of Public Instruction, on the extract of the verbal

Art. 2

Notre Ministre Secrétaire d'État au département de l'Instruction publique ~~et des Cultes~~ est chargé de l'exécution du présent décret.

Fait au Palais des Tuileries le 25 Janvier 1865

Signé : Napoléon.

Par l'Empereur :
Le Ministre Secrétaire d'État
au département de l'Instruction publique ~~et des Cultes~~,
Signé   V. Duruy
Pour ampliation :
Le Secrétaire général,

process of the session of 23 January 1865 of the Academy of Sciences of the Imperial Institute of France, we decreed and decree that which follows: First article—

The election by the Academy of Sciences of the Imperial Institute of France of M. Foucault to fill the seat of academician that became vacant in the mechanics section, following the death of M. Clapeyron, is approved. . . .[76]

Foucault's ultimate acceptance by the group of theoreticians that dominated the French Academy of Sciences signaled the final vindication of the work of a man whom many of these people considered untrained and unsophisticated in mathematical theory. This was a triumph for science, and it announced that not only theoreticians could practice it effectively, but rather that other people—those who possess a keen sense and intuition about nature—could contribute much to our knowledge. It took an intervention by an emperor to push this exclusive group of scholars to admit an untrained scientist into their midst; but it was done—and science was richer for it.

Foucault's pendulum was again displayed in the next Universal Exhibition in Paris, which took place in 1867. The historian Pierre de La Gorce described the exhibition as follows.

> Of all the exhibitions, that of 1867 remains memorable for two reasons; first by the display of magnificence which no one had dreamt of before; in the second place by the violent gusts of uneasiness which, cutting across public gaiety, came near to dispersing it all more than once.[77]

The Second Empire was near its peak by this time, with its gaudy displays and relentless pursuit of pleasure and national glory. Across the border, the North German Confederation had formed, and all Europe could feel the potential danger of the strengthening of a giant that was to menace Europe for three-quarters of a century to come. Louis-Napoléon was suffering from persistent bladder problems, which left him in pain and made it hard for him to walk or ride a horse. The economy was weakening, and an opposition to the regime was slowly emerging. For the first time, the emperor was not making decisions on his own but was increasingly relying on Empress Eugénie and on his close advisers. They all felt that a grand display would be a welcome diversion for the people from the empire's emerging economic and political problems. Baron Haussmann organized the greatest public spectacle the French capital had ever seen.

The entire project was a tremendous engineering feat. The grounds of the Champ de Mars were covered by a layer of soil; special steam pumps were installed at the Trocadero area to pump up water from the Seine to a height of 135 feet for the fountains, and to ensure sufficient pressure, two windmills were constructed there; a large palace resembling the Roman Coliseum was built to include six large galleries for exhibits of different countries. The structures were painted brown and gold. Pavements made of compressed concrete connected the palace with the rest of the display areas. When everything was in place, one could walk into a complete Tyrolean village; an Egyptian caravanserai surrounded by minarets; a Hungarian settlement; an American ranch; and Japanese, Mexican, and Russian compounds. In all, 52,000 exhibitors were present.

While Haussmann was the architect of the exhibition, the grand exhibit of French machines and instruments was under the charge of Léon Foucault. With great care and his typical worries about every detail, Foucault arranged the display of steam engines, astronomical clocks, electric light regulators, and electric machines of various kinds. Some of the many items displayed were of his own invention.

Louis-Napoléon and Eugénie inaugurated the exhibition. Next to them on stage stood Foucault and other members of the organizing committee. The French national anthem was sung and played on an organ, and when it ended, all the machines on display were turned on: steam engines and electric motors all moved in unison to the great amazement of the crowd. Foucault's pendulum swung back and forth in full view of the assembled crowds.

This was a high point in Foucault's career. He was finally a member of the Academy of Sciences, a recognized scientist, and beloved by the emperor who had bestowed on him great honors. It was a high point for the emperor and the empress, who showcased the greatness of France to the world. Few would have guessed that within a short time, both Foucault and Louis-Napoléon would be gone; and so would the empire.

16

———◆———

A PREMATURE END

Three months after the inauguration of the Universal Exhibition, on July 10, 1867, Foucault began to feel the first signs of paralysis. It started with numbness in his right hand, which made it difficult to sign his name.

In the charming pavilion on the rue d'Assas, which one day he would have inherited from his mother, everything had been prepared for Foucault so that he could pursue his work in comfort. His mother made all the arrangements, even adding a large balcony to the apartment to house the *siderostat,* Foucault's latest invention. The design of the siderostat was inspired by the heliostat, an instrument for looking at the Sun. When Foucault was getting weaker, he thought about finding a way to continue his work at home. The siderostat allowed him to project an image of a section of the sky, which he could observe even while lying on his couch at home.

Foucault was becoming sicker, and Mme. Foucault went to great lengths to assure her son's comfort. His study was decorated in calming, warm colors, and the walls were lined with soft material designed to absorb sound. Foucault was no longer able to walk to the Observatory, just a few short city blocks away. His paralysis had progressed quickly, making movement difficult.

At home, Foucault was often surrounded by the many friends who came to offer support and talk about science, inquiring about his latest ideas. Foucault would discuss these at length and add that he had so many new ideas that it would, indeed, take twenty years to see them all to fruition. But soon his condition became even worse, making him lose the use of his right hand. He felt lost, but kept his pain to himself, continuing to meet his friends and associates and well-wishers. He had more friends now than ever before. He even tried to continue working on his myriad projects in astronomy and physics. But the terrible illness continued its advance. He lost the use of his tongue and then his vision. His intelligence remained intact throughout, however, making the loss of his abilities even more painful. When a friend once tried to interpret his thoughts and put them in words for him, Foucault lit up for a moment: "That's it," he said, "that's it." Then he sank into despair.

We don't know exactly what Foucault suffered from, but it seems to have been Lou Gehrig's disease, or a similar degenerative illness. His handling of toxic chemicals and dangerous metals such as mercury over extended periods of his life may have had something to do with his condition. Foucault suffered for six months. Then, on February 11, 1868, he died. He was forty-nine years old.

Louis-Napoléon was saddened by the loss of his favorite physicist. He, along with Foucault's many friends, mourned his passing. The emperor did not want to see Foucault's work and great achievements lost or forgotten. He also wanted to encourage scientists to pursue further the research projects begun by Foucault. To achieve these goals, Napoléon III summoned a committee made up of Foucault's closest friends within the scientific community, as well as leaders of industry. The committee included M. Roland, Director General of State Manufacturing; Professor of Medicine Jules Regnault of the Central Pharmacy; Professor Wolf of the Paris Observatory; Professor Adolphe Martin; and Jules-Antoine Lissajous, Professor at the Lycée Saint-Louis and member of the Academy of Sciences.[78]

Based on the recommendations of the committee, Napoléon III set aside an annual sum of 10,000 francs for the purpose of publishing and promoting Foucault's works and ideas. In thanking the emperor for his generosity to the memory of Foucault, Lissajous wrote:

Thanks to the union of strong goodwill and friendship, thanks to the personal liberal feelings of the sovereign, and with the support of his mother and the family of Léon

Foucault, this important mission will be fulfilled and it will safeguard the scientific heritage of a man whose discoveries have enriched science and brought honor to his nation. The Society of Friends of the Sciences permits us to pay our respectful and sincere homage to the benevolent interest shown by the Emperor in all the important research projects, to his profound generosity, which is rendered spontaneously to scholars without in any way hampering their liberty.[79]

These were interesting words. It appears from the passage that at least some of the scientists of the Second Empire did not feel censored or in any way restricted in their work and felt they had preserved their academic freedoms. These feelings are in sharp contrast with the state of journalism, which was regularly censored during the Second Empire, as well as the condition of the legislative and judicial bodies—all tightly controlled by the emperor—and the fact that their members were no longer elected by the people. Louis-Napoléon was true to the cause of science to the very end. And his dedication to Foucault and his legacy was complete.

Unfortunately, the emperor's goodwill was not enough. For Louis-Napoléon himself was to meet his demise before the project to commemorate the works of Foucault could come to fruition. It is moving that Louis-Napoléon was willing to spend so much money on immortalizing the physicist at a time when money was in short supply in France. The Second Empire was clearly in decline in the late 1860s, the economic boom of the earlier years having run out of steam. The emperor was even hard-put to find money to support the Imperial Observatory he had cared so much about. A letter sent during those years by Le Verrier directly to the emperor pleads urgently for money to support the operations of the Observatory. And Napoléon III had many, many other worries on his hands during those difficult years. For once the economic decline had begun, the empire was ready to collapse. The final demise would be a military one, at the hand of the Prussians.

17

THE DEFEAT AT SEDAN

The Second Empire enjoyed one last spark of glory in November 1869, with the opening of the Suez Canal. Built by French engineers, the canal was a marvel of technology, and it remains an immensely important waterway for global trade and commerce. The Empress Eugénie presided over the ceremonies inaugurating the canal, with an impressive display of warships of the French fleet sailing down the waterway. This was the last great public spectacle of the Second Empire. Soon, trouble began. It was brewing in a faraway location, on the other side of the Mediterranean from the Suez Canal—in Spain.

The Spanish had just deposed their queen, the corrupt Isabella II, and wanted a new monarch. They began a search among the royal families of Europe. One of their prime candidates was young Prince Leopold von Hohenzollern, a member of the Prussian royal family. When this was known around Europe, the French became alarmed. They worried about the possibility of having *two* Prussian rulers on their borders: one to their northeast and one to their south.

Fortunately, Leopold lacked the ambition to become the king of Spain. He viewed the Spanish throne as unstable and anyway preferred the life of an idle prince. His uncle, Wilhelm II of Prussia, agreed and was content to let Leopold pass up this opportunity. But Prussia's minister-president, Otto von Bismarck, saw here an opportunity to humiliate France. Eighteen months after Leopold's rejection of the Spanish throne, he urged the prince to change his mind and become the king of Spain. Bismarck used his influence with Leopold's father, conspired with the Spanish parliament, and pressured King Wilhelm II. Finally, he was successful, and Leopold reluctantly agreed to accept the Spanish throne.

As Bismarck had expected, the French were furious. And they had every right to be. According to tradition, the negotiations to fill vacant European thrones were conducted in the open, so that every country could express its concerns about each candidate, and in a way that allowed the final choice to be acceptable to all. In this case, a bellicose outsider was intervening with the purpose of antagonizing France and hurting Napoléon III.

Louis-Napoléon's health was deteriorating further. His bladder problems were causing him much pain, and he was often immobilized. Several doctors examined the emperor, but none could determine the root of the problem. One of them suggested exploratory surgery, but this was never carried out. The weakened emperor met with his ministers and empress trying to decide what to do. He now felt too weak to make state decisions on his own.

Eugénie and some of his cabinet members were leaning toward war with Prussia. Eventually, however, France applied diplomatic pressure, and the Prussians reneged. Leopold withdrew his candidacy for the Spanish throne, and it seemed that the conflict was averted.

But by then the French public was already clamoring for war. The French sent their ambassador to speak to King Wilhelm and demand that he give them a written guarantee that such a situation would never occur again. Wilhelm refused to sign such a humiliating paper. He felt that Leopold's withdrawal should have been enough for the French and that nothing more was necessary. Now nothing could prevent a war between France and Prussia.

The science-loving emperor set to work to prepare his country for war. The French developed a new rifle and invested in arms and supplies to bring their army up to the standard they felt was necessary to match the Prussian war machine. Louis-Napoléon, frail and ill as he was, decided to lead his army into battle. This was the Napoléonic way, he knew. His uncle had always led his forces into battle, regardless of what the outcome might be. People close to the emperor knew that the bladder problems he had been suffering from made it very painful for him to sit on a horse, but there was no talking the emperor out of his decision to be at the head of the French army. Despite all the preparations, the French were not ready for war. French generals knew quite well that the Prussians were far better prepared for battle.

In July 1870, Louis-Napoléon arrived at Metz, a city in northeast France near the German border, to take command of his Army of the Rhine facing the Prussians. He found the army in a bad state of preparedness, with only 200,000 men, half the number that should have been recruited by that time. There was chaos everywhere, with commanders unable to find their troops and with communication disruptions. Morale was low.

The war began with a minor victory for the French, who managed to take a small town on the other side of the Rhine. But very soon the tide turned against them. The Prussians unleashed deadly attacks against the Army of the Rhine and after a few local victories managed to split the French army in two. The retreating French left the front wide open, and the Prussians were facing the undefended central part of France. As they forged ahead, the French army retreated and eventually regrouped near the pastoral town of Sedan, close to the Ardennes and the Belgian border. At 4 A.M. on September 1, 1870, German forces numbering over a quarter-of-a-million men surrounding 100,000 French troops began their attack. Even before the battle started, it was clear to everyone there that the French army had no chance at all. The battle of Sedan became a massacre of the French army, and the emperor, who survived the attack, was taken into Bismarck's headquarters as prisoner of war.

The Prussians pressed on toward Paris, and their artillery bombarded the French capital for several days before an armistice was called. The Prussians demanded and got Alsace and Lorraine as part of their spoils of war. Louis-Napoléon was kept prisoner for some time at the castle of Wilhelmshohe, where he was more comfortable than he had been during his captivity, years earlier, at the Ham. By coincidence, Wilhelmshohe had been the residence of Louis-Napoléon's uncle, King Jerome of Westphalia, during the first empire. And the palace contained pictures of Louis-Napoléon's mother, Hortense. After the terms of peace had been agreed upon, in March 1871, the deposed emperor was allowed to leave and join his wife and child in exile in England. The Second Empire had come to an end.

The demise of Napoléon III prevented the project of commemorating Foucault's life's work from becoming a reality. The funding was no longer there, and the former emperor had other problems to worry about in his new life in exile. Two years after his arrival in Britain, on January 9, 1873, Louis-Napoléon Bonaparte died following a bladder operation. Eugénie survived him to live another forty-seven years, become a close friend of Queen Victoria, live through World War I, and die in Madrid in 1920.

But the group of Léon Foucault's friends, which the emperor had convened during happier times, remained active and continued their efforts to preserve his legacy. These friends remained in contact with his mother. Mme. Foucault, eager to see the project of the publication of her son's works to its fruition, helped his friends and contributed the money they needed. The collected works of the physicist were finally published in Paris in 1878, ten years after his death.

18

———◆———

AFTERMATH

After Foucault's death, more experiments with his pendulum were carried out around the world, partly to check whether the sine law indeed worked as expected and partly because these demonstrations were so enjoyable as well as educational for the public. One test was again done in Holland, in Haarlem; another again in France, north of Paris in the region of Picardy, in the medieval city of Amiens. The beautiful cathedral in this city was built in 1218, on the site of a Roman church dating as early as 354 A.D. The church provided an impressive setting for the experiment.

But the most magnificent demonstration of Foucault's pendulum had taken place much earlier, in 1852, in the very high cathedral in the German city of Cologne. The line used for the pendulum in this display was over 160 meters (over 500 feet) long. This made the swing of the heavy pendulum bob very slow and dramatic. This experiment was carried out by C. Garthe, who was so careful that he launched his pendulum from a wide variety of initial directions to make sure that the results were accurate and consistent.

There had been tests in Groningen and in Deventer in Holland, then in the city of Danzig on the Baltic Sea. In 1854, the Foucault experiment was once more repeated in the New World, this time in Saint Gall, Quebec. In 1895, an experiment was carried out near Copenhagen. The results of all the tests accorded with Foucault's sine law.

Then, in 1902, one of the most important experiments was done—it was a replay of the original 1851 demonstration in the Panthéon. This happened after the Panthéon had again become a shrine to the great citizens of France following a period during which it had served, again, as a church. In this new period of secularism, the Astronomical Society of France decided to commemorate Foucault's Panthéon demonstration in the public way it was done over half a century earlier. In the meeting the society held on January 8, M. Fontvielle said that he hoped that the well-known astronomer Camille Flammarion (1842–1925) would agree to intercede with the authorities on behalf of the Society to allow the new commemorative experiment to be conducted in the Panthéon. The February 5 meeting notes recall that in the previous meeting "the Society warmly endorsed the view expressed by M. Fontvielle that it would be desirable to see under the Panthéon's dome Foucault's beautiful and instructive experiment (interrupted by the coup d'état of December 1851 before one could have obtained from it all the conclusions that could have been expected)." The members agreed to push forward with the project after the centenary of Victor Hugo's birth, which was to take place in the Panthéon on February 26. The President of the Astronomical Society of France at that time was the French mathematician Jules Henri Poincaré (1854–1912), one of the greatest theoretical mathematicians of all time. The fact that a mathematician of his stature was so enthusiastically involved with reviving Foucault's experiment shows just how far Foucault's reputation had come, and that—long after his death—he was finally accepted by this profession.

As soon as they received permission to go ahead with the project, the architect working on the Panthéon was consulted on the issues of constructing the support for the pendulum. Camille Flammarion began to work on the pendulum project together with an engineer hired for this grandiose reconstruction of the experiment, Alphonse Berget. The pair tried to get Foucault's original pendulum, but the museum to which Foucault had willed the pendulum refused to part with it. As a substitute, Flammarion and Berget used the pendulum that Maumené had used in the Reims Cathedral in 1851. But they were able to procure the other elements of the original display, including the elegant mahogany balustrade used by Foucault.

When Flammarion and Berget tested their equipment at the end of June, they found that their pendulum was almost a meter too short. They had to quickly solve the problem since the public

The 1902 pendulum experiment in the Panthéon by Camille Flammarion (bearded) and Alphonse Berget (with glasses). The Minister of Public Instruction burns the line holding the bob.

display was scheduled to begin in early July. The pair scrambled to find a solution; finally they decided to raise the wooden platform with the sand and displayed numbers so it would reach the bottom of the pendulum.

In July, Flammarion began to demonstrate the pendulum to groups of secondary school children—he valued the educational role of Foucault's experiment very much. Then on October 22, the big public demonstrations of Foucault's pendulum began. The first started at 2 P.M., with over 2,000 people present inside the Panthéon. Among the guests were the composer Camille Saint-Saëns (1835–1921) and the sculptors Auguste Rodin (1840–1917) and Frederic Auguste Bartholdi (1834–1904)—the latter the creator of the Statue of Liberty. The French Minister of Public Instruction, M. Chaumié, started the pendulum in motion, as had been done in earlier demonstrations, by putting a match to the thin thread holding the pendulum's bob to the side. There were public speeches, and Flammarion gave homage to Foucault:

> The most magnificent lesson in popular astronomy ever given to the public was surely the memorable experiment conducted at this very place half a century ago by Léon Foucault. It was a practical, evident, and majestic demonstration of the movement of rotation of our globe and a grammatical affirmation of the title planet, or "moving star," to the world in which we live.[80]

"The image of Galileo has just passed before our eyes," he said, as the pendulum swung in front of the audience. He continued to recall the age-old controversy between belief and science and said: "The doctrine of the Earth's movement has changed philosophy as a whole. It is the greatest moral and ethical revolution in the history of mankind."

Over the next year, until the pendulum was taken down again, tens of thousands of people flocked to the Panthéon, arriving at any hour of the day, to witness Foucault's triumph and hear scientific explanations by Flammarion or Berget about how and why the pendulum swung the way it did and its implications.

Foucault's great triumph is a triumph of the human mind. It is a double victory of knowledge against ignorance. First, Foucault's great achievement showed how physical intuition, engineering skills, and perseverance can win against the hubris of mathematics detached from the real world. More important, Foucault's landmark experiment spelled the end of speculations and persistent false beliefs. As such, Foucault's definitive proof of the rotation of the Earth helped vindicate Galileo, Copernicus, and Giordano Bruno. After Foucault's successful demonstration of the Earth's rotation, Church scholars themselves embraced the heliocentric, Copernican view of the world, and openly wrote about Foucault's proof.

In 1911, the Jesuit Father J. G. Hagen wrote a major treatise called *The Rotation of the Earth: Its Mechanical Proofs Ancient and New.*[81] Hagen described in detail the Foucault pendulum experiment carried out in Rome in 1851, as well as other experiments that followed. He explained Foucault's sine law and how it works, giving all the scientific details of the proof of the rotation of the Earth.

The fact that a Jesuit priest should write a book, published by the Vatican, describing the rotation of the Earth, goes to show how far the Church had come in accepting science. Hagen's writings make it clear that it was, in fact, the discovery made by Foucault, and nothing else, that finally convinced the Church that the Earth rotates. And finally, Pope John Paul II's "apology to Galileo" in October 1992 owes everything to the work of Léon Foucault almost a century and a half earlier.

Foucault received more honors after his death. His name was etched into the metal structure of the Eiffel Tower, along with the names of France's greatest citizens. A statue of Foucault was erected and now adorns one of the façades of the Paris City Hall—the Hôtel de Ville. And when we look at the moon through a telescope, we may see a crater, 22 kilometers in diameter, located to the northwest of the Sea of Rains *(Mare Imbrium)*, at 40 degrees west and 50 degrees north. This crater is named Léon Foucault.

APPENDIX

◆

PROOFS OF FOUCAULT'S SINE LAW

There are many proofs of Foucault's sine law: some are detailed and complicated, others simpler. Here we give two proofs. The first, the proof due to Liouville, is the simplest one available, but it lacks detail and thus assumes much on faith. It is a "mathematician's proof." The second proof is somewhat more detailed, but not too difficult.

LIOUVILLE'S PROOF OF FOUCAULT'S SINE LAW

The simplest and most graphic proof was proposed by Liouville shortly after Foucault presented his law. The proof is given below.

The rate at which our planet rotates around the *vertical* at any given point on Earth is the pertinent parameter in this situation, since this is the rotation that Foucault's pendulum "measures" by its immobility and our rotation right under it. At the pole, rotation about the vertical is indeed the full rotation of the Earth. On the equator, on the other hand, the rotation about the vertical is zero and the pendulum's plane of oscillation doesn't change at all.

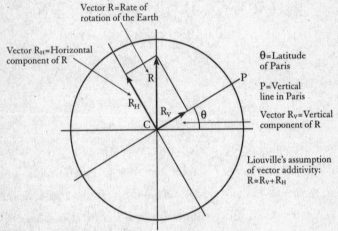

Vector R=Rate of rotation of the Earth

Vector R_H=Horizontal component of R

θ=Latitude of Paris

P=Vertical line in Paris

Vector R_V=Vertical component of R

Liouville's assumption of vector additivity: $R=R_V+R_H$

Foucault's pendulum in Paris is affected only by the vertical component of the rotation, R_V. From simple trigonometry: $R_V=R \cdot \sin\theta$. Hence the time to complete a turn at latitude θ: $T=\frac{24}{\sin\theta}$

Liouville's idea is that of simple *vector addition*. The rate of rotation, R, viewed as a vector, is equal to the sum of two vectors: R_V, rotation about the vertical; and R_H, rotation about a line perpendicular to it and aligned with the local meridian. Liouville's vector equation is: $R=R_V+R_H$. This is shown in the figure above.

Using trigonometry, we see that $Rv=R\sin(\theta)$, where θ=latitude of the observer.

Liouville's proof is visually clear, but it lacks both mathematical and physical detail. Most importantly, it assumes—on faith, as it were—that these vectors are indeed additive. This is a mathematician's proof; it makes perfect sense if one already believes that the sine law is correct, but it probably would not convince the nonbeliever. Liouville was criticized for his proof's lack of detail as soon as he proposed it, but still he was the first to provide some proof of Foucault's law. His proof, therefore, has historical value. A more complete proof (one of many that have been proposed since Foucault first derived the sine law) follows.

AN ALTERNATIVE, DETAILED PROOF OF
FOUCAULT'S SINE LAW

The figure below shows the circle of sand on the ground at the Panthéon, the latitude, the north pole and the equator, and the angles relevant to the pendulum.[82]

Consider the relative speeds of motion of the extreme north and the extreme south points of the circle of sand of radius r as shown in the figure. The south point is farther from the axis of rotation of the Earth. It will therefore move *faster* than the north point. Let ω denote the angular velocity of the Earth. Let R be the radius of the Earth. From the figure, and using elementary trigonometry, we see that the center of the circle of sand moves

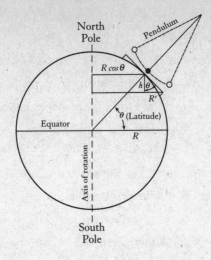

with velocity $\omega R \cos\theta$, where θ = the latitude of Paris (48 degrees, 51 minutes North).

Now, the northernmost point of the circle moves with velocity

$$\omega R \cos\theta - \omega r \sin\theta$$

as we see from the figure using trigonometry. Similarly, the southernmost point of the circle moves with velocity

$$\omega R \cos\theta + \omega r \sin\theta$$

The difference between the velocity of either of these two points and that of the center of the circle is:

$$\omega r \sin\theta$$

If the pendulum starts its motion in the north-south plane, the east-west component of its velocity will be the same as that of the center of the circle. The circumference of the circle, as we know, is equal to $2\pi r$. Thus the time for a full circuit is given by:

$$T = 2\pi r/\omega r \sin\theta = 24 \text{ hours}/\sin\theta$$

because $\omega = 2\pi/24$ hours, since the Earth rotates a full circle in twenty-four hours.

QED

ACKNOWLEDGMENTS

———◆———

In the summer of 2001, while in Paris to interview the eminent French physicist Alain Aspect, I visited the Panthéon. I was struck by the monumental beauty of Foucault's pendulum, slowly swinging beneath the great vault, just as it did when Foucault first demonstrated it 150 years earlier. (Regrettably, the following year the French authorities removed the pendulum to make room for an exhibition by a minor German artist.)

When, a few months later, I described the pendulum and Foucault's experiment to my friend and editor, Tracy Behar, she suggested that I write a book about it. I thank Tracy for this idea and for her encouragement and support throughout the process of writing this book.

I returned to Paris in 2002 to research and write the book. Many people, in Paris and elsewhere, were very helpful to me in this project, and I thank them all for their unselfish contributions to this effort.

I am most grateful to my friend the renowned French historian Evelyne Patlagean. Evelyne helped me navigate the complex and deeply rewarding labyrinth that is intellectual Paris. She directed me to innumerable historical sources pertinent to the story of the pendulum, and our discussions improved my understanding of France, its people, and its rich history.

I thank Evelyne's daughter, my friend the mathematician Marina Ville, for discussions of the mathematics of Foucault's pendulum and for steering me to many important sources and experts on the history of science and nineteenth-century France.

I want to express my gratitude to my good friend the mathematician Carlo F. Barenghi for coming to visit me in Paris to discuss Foucault and his work as well as the mysterious force of Coriolis, which we explored experimentally on every merry-go-round in the French capital with our daughters.

I thank my friend the publisher Dominique Carré for help in researching my story of Foucault's pendulum and for giving me a copy of a book he published in 1990, Stéphane Deligeorges's excellent biography of Foucault.

Many thanks to my research assistant, Robin Frans, for much help in the early stages of researching this project. Robin has uncovered numerous sources of information on the pendulum that were not easy to obtain.

I am grateful to Michael Johnson of the University of Illinois for a discussion of the Second Empire.

I thank Guy Cassegrain, Françoise Launay of the Paris Observatory, and André Baranne of the Marseille Observatory for information on the astronomer Laurent Cassegrain.

Much of the material on Foucault's life and work comes from original nineteenth-century sources found in the wonderful archives and libraries of the French capital.

I am grateful to the Institut de France for permission to use its substantial archives. I thank the Perpetual Secretary of the French Academy of Sciences, Jean Dercourt, for welcoming me as a guest to the Institute; I am most grateful to Mireille Pastoureau for allowing me access to the archives of the Institute. I thank Annie Chassagne for permission to reproduce documents.

I appreciate the kindness of the French Academy of Sciences for allowing me access to its information and documents relating to Foucault and his work. I am grateful to Marie-Josèphe Mine, Florence Greffe, and Claudine Pouret for producing documents and granting me permission to reproduce them. I thank Suzanne Nagy for photographing various documents.

No research on any topic in France would be complete without consulting at least some material in France's colossal National Library, the Bibliothèque nationale de France. I thank the librarians of the BnF for their generosity with their time and their help in sifting through a large volume of original material.

Many thanks to the administration and staff of the Bibliothèque Historique de la Ville de Paris for providing me with many historical documents on the Second Empire, Napoléon III, and Foucault's experiments.

I am grateful to the Conservatoire National des Arts et Métiers (CNAM) in Paris for making its library available to me, as well as its wealth of artifacts relevant to the work of Foucault. Many thanks to CNAM for permission to reprint pictures of Foucault, the original 1851 pendulum used in the Panthéon demonstration, and the daguerreotype of the Sun. I thank Thierry Lalande, Isabella Feral, and Frederique Desvergnes at CNAM for all their help.

I am grateful to the Observatoire de Paris for allowing me access to its library and Meridian Hall (Salle Cassini), the site of Foucault's first public display of his pendulum. At the observatory, I thank the librarian Josette Alexandre and Nino Azzi for their help.

I am most grateful to André Thiot of the Société Astronomique de France (SAF) for welcoming me at the Society's offices and library and for invaluable help in my research. Thanks also to Gean Renard at SAF, and other members, enthusiastic French astronomers, for their help.

I thank Scott Steedman of Raincoast Books in Vancouver for several good ideas about the Second Empire, which improved the book.

In addition to my editor, Tracy Behar, I thank the many dedicated editors and staff at Atria Books. I am grateful to Wendy Walker for her help in editing the manuscript and obtaining the figures. I thank Brenda Copeland for all her help. I thank copy editor Red Wassenich for his wonderful editing of the manuscript.

Finally, I am most grateful for my wife, Debra, for her many ideas that greatly improved this book, for reading the manuscript a number of times, and for countless conversations we have had about Foucault.

NOTES

———◆———

[1] Joshua 10, 12–13.

[2] Isaiah 38, 7–8.

[3] Ecclesiastes 1, 5.

[4] Philolaus, *Treatise of the Sky,* Book II, 13, 293. Reported in Gapaillard (1993).

[5] It should be noted that these ellipses—in our solar system—are close to circles. In some extra-solar systems (systems of planets found around other stars) the eccentricity of the orbits is higher.

[6] Today we know that this law is an approximation; it is accurate only when the oscillations are not too large. The differential equation governing the general movement of the pendulum cannot be solved exactly.

[7] Jacques Gapaillard, *Et pourtant elle tourne! Le mouvement de la Terre.* Paris: Éditions du Seuil, 1993, p. 32.

[8] Académie des Sciences, *Comptes rendus hebdomedaires de l'Academie des Sciences,* April 4, 1851.

[9] A. Koyré, *Chute des corps et mouvement de la Terre, de Kepler à Newton.* Paris: Vrin, 1973. The famous French astronomer Camille Flammarion also mentioned the cannon story in his address given at the 1902 demonstration of Foucault's pendulum in Paris. According to Flammarion's version of the story, the monk and his assistant fired three cannonballs. The first one disappeared,

the second fell 2,000 feet to the west, and the third 2,000 feet to the east. At that moment, the pair decided that by the law of averages, the next cannonball should fall right on their heads and immediately stopped the experiment.

[10] Mersenne was the discoverer of the Mersenne prime numbers (not all of which, unfortunately, are real primes). There is a worldwide search for such numbers, carried out on the Web, under the name of the Great Internet Mersenne Prime Search (GIMPS).

[11] *Correspondance du P. Marin Mersenne, religieux minime.* Cornelius de Waard, ed. Paris: Editions du Centre National de la Recherche Scientifique, 1969. Author's translation.

[12] E. T. Bell, *Men of Mathematics: The Lives and Achievements of the Great Mathematicians from Zeno to Poincaré.* New York: Simon & Schuster, 1965, p. 93.

[13] Steven Weinberg, *Gravitation and Cosmology: Principles and Applications of the General Theory of Relativity.* New York: Wiley, 1972, p. 13.

[14] Jacques Gapaillard, *Le mouvement de la Terre: la détection de sa rotation par la chute des corps.* Paris: Cahiers d'histoire et de philosophie des sciences, No. 25, 1988.

[15] Reprinted in Jacques Gapaillard, *Et pourtant elle tourne! Le mouvement de la Terre.* Paris: Éditions du Seuil, 1993, pp. 131–2. Author's translation.

[16] Joseph Bertrand, *Éloge historique de Léon Foucault.* Paris: Institut de France, 1882, p. 18. Author's translation.

[17] Jacques Gapaillard, *Et pourtant elle tourne! Le mouvement de la Terre.* Paris: Éditions du Seuil, 1993, p. 235.

[18] Stéphane Deligeorges, *Foucault et ses pendules.* Paris: Editions Carré, 1990, p. 48.

[19] Jules-Antoine Lissajous, *Notice historique sur la vie et les travaux de Léon Foucault (de l'Institute).* Paris: P. Dupont, 1875.

[20] David Boyle, *Impressionist Art,* Vancouver: Raincoast, 2001, p. 28.

[21] Alfred Donné, *Cours de microscopie.* Paris: Baillère, 1845.

[22] Joseph Bertrand, *Éloge historique de Léon Foucault.* Paris: Institut de France, 1882, p. 3. Author's translation.

[23] Joseph Bertrand, *Éloge historique de Léon Foucault.* Paris: Institut de France, 1882, p. 4. Author's translation.

[24] Joseph Bertrand, *Éloge historique de Léon Foucault.* Paris: Institut de France, 1882, p. 5. Author's translation.

[25] Adapted from François Sarda, *Les Arago: François et les autres.* Paris: Tallandier, 2002, pp. 70–76.

[26] This problem has a long history and the interested reader is referred to books on the history of quantum mechanics for a complete discussion.

[27] Jules-Antoine Lissajous, *Notice historique sur la vie et les travaux de Léon Foucault (de l'Institute)*. Paris: P. Dupont, 1875.

[28] It should be noted that other experiments provide equally definitive evidence for the particle theory of light—in particular, Albert Einstein's 1905 work on the photoelectric effect, for which he won the Nobel Prize.

[29] Joseph Bertrand, *Éloge historique de Léon Foucault*. Paris: Institut de France, 1882, p. 11. Author's translation.

[30] Joseph Bertrand, *Éloge historique de Léon Foucault*. Paris: Institut de France, 1882, p. 7. Author's translation.

[31] The CNAM is housed in what used to be, until 1799, the priory of the medieval church of Saint-Martin des Champs, built in the eleventh century. The Foucault pendulum in this converted ancient church impressed Umberto Eco so much that he decided to use it as the setting for the opening of his novel *Foucault's Pendulum,* inspiring his book's title.

[32] Stéphane Deligeorges, *Foucault et ses pendules*. Paris: Éditions Carré, 1990, p.59.

[33] This story is told in an upcoming book by Simon Winchester, *Fatal Equation.*

[34] Latitude 48 degrees and 51 minutes means that θ=48.85 degrees (since 51/60=0.85). Using a calculator, we find: sin (48.85)=0.753. Now we compute: T=24/sin (θ)=24/0.753=31.9 hours. Try this with the latitude of your own location. You can find your latitude by consulting any map of your area. For U.S. addresses, the exact latitude (even as accurate as the location of many public buildings and landmarks, rather than generally for an entire city or town) can be found at the U.S. Geological Survey Web site http://geonames.usgs.gov.

[35] There are several different proofs of the sine law; two such proofs are given in the appendix.

[36] Académie des Sciences, *Comptes rendus hebdomadaires de l'Académie des Sciences,* February 10, 1851, pp. 157–8. Author's translation.

[37] Astronomers call the exact time for rotation computed from positions of stars a *sidereal* day, and it is slightly shorter than the usual (solar) day of twenty-four hours, but the difference is automatically corrected on the autumnal equinox. As the Earth rotates about its axis, it also orbits the Sun, and therefore, by the end of each day, it has progressed some distance around the Sun, so that the sidereal day is about four minutes shorter. After the Earth has turned far enough for any star to return to its apparent position in the sky, the Earth must turn an additional 1/365 of twenty-four hours (which is equal to 3 minutes, 56.71 seconds) for the Sun to return to the local meridian. Thus a solar day is a little under four minutes longer than a sidereal day. But this difference gets automatically readjusted during the autumnal equinox, at which time the sidereal

day and the solar day coincide. (For more on this topic, see Jay Pasachoff, *Astronomy: From the Earth to the Universe*, in the references. The distinction is not important for what follows.) To see that the statement "The angular momentum is thus 15" x sin(γ) for a second of sidereal time (being 15 degrees in a sidereal hour)" is equivalent to the sine law as given earlier in this book, note the following. The angular momentum is the rate of rotation of the Earth given in degrees per hour. That rate is 15 degrees per hour. The reason for this answer is the simple calculation that the Earth goes around full circle, meaning 360 degrees, in twenty-four hours. Thus in one hour, the Earth completes 1/24th of a full circle, and hence it rotates by 360/24=15 degrees. How long does it take Foucault's pendulum to go around full circle? At the pole, where the rate of rotation is 15 degrees per hour, it takes twenty-four hours. At a location with latitude θ, it should take longer by a factor inversely related to the rate of rotation (since time=distance/speed). By Foucault's sine law as given in this book, we know that we must divide the time (twenty-four hours) by the factor sin (θ). The speed (angular momentum) must be affected by latitude inversely to the way time is affected by latitude (because speed and time have an inverse relationship). Therefore, Binet adjusts the angular momentum by *multiplication* by sin (θ). Note that this is not a proof of Foucault's sine law, which is much more involved. This is only an argument that shows that Foucault and Binet were talking about the same thing. Finally, note that there are 3,600 seconds in an hour, and 3,600 seconds in a degree (60 minutes times 60 seconds) and hence Binet's parenthetical remark about the sidereal hour.

[38] Jacques Gapaillard, *Et pourtant elle tourne! Le mouvement de la Terre.* Paris: Éditions du Seuil, 1993, p. 250.

[39] Académie des Sciences, *Comptes rendus hebdomadaires de l'Académie des Sciences,* February 3, 1851, pp. 135–6. Author's translation.

[40] Jean Plana. *Note sur l'expérience communiquée par M. Léon Foucault.* Paris: Academie des Sciences, 1851. Author's translation.

[41] Joseph Bertrand, *Éloge historique de Léon Foucault.* Paris: Institut de France, 1882, p. 21. Author's translation.

[42] John Bierman, *Napoléon III and His Carnival Empire.* New York: St. Martin's Press, 1988, p. xiii.

[43] John Bierman, *Napoléon III and His Carnival Empire.* New York: St. Martin's Press, 1988, p. 41.

[44] From Bibliothèque Historique de la Ville de Paris.

[45] John Bierman, *Napoléon III and His Carnival Empire,* New York: St. Martin's Press, 1993, p. 60.

[46] Jasper Ridley, *Napoléon III and Eugénie.* New York: Viking, 1979, p. 238.

[47] Jacques Gapaillard, *Et pourtant elle tourne! Le mouvement de la Terre.* Paris: Éditions du Seuil, 1993, p. 242.

[48] Joseph Bertrand, *Éloge historique de Léon Foucault.* Paris: Institut de France, 1882.

[49] *Journal des Débats,* Saturday, January 4, 1851.

[50] Philippe-Louis Gilbert, *Léon Foucault, His Life and Scientific Work.* Brussels: A. Vromant, 1879.

[51] Reported in Stéphane Deligeorges, *The Foucault Pendulum in the Panthéon.* Paris: Musee du Conservatoire National des Arts et Métiers, 1997, p. 4.

[52] Stéphane Deligeorges, *The Foucault Pendulum in the Panthéon.* Paris: Musée du Conservatoire National des Arts et Métiers, 1997, p. 9.

[53] *Journal des Débats,* Monday, March 31, 1851, p. 4.

[54] *Journal des Débats,* Monday, March 31, 1851, p. 4.

[55] Stéphane Deligeorges, *Foucault, et ses pendules.* Paris: Editions Carré, 1990, p. 65.

[56] On October 24, 1887, a demonstration of Foucault's pendulum took place at the St. Jacques tower in Paris. W. De Fontveille published a commentary on it in a publication called *French Expeditions to Tonkin.* The event was attended by the Emperor of Brazil and the Chinese general Tcheng Ki-Tong. In his commentary, Fontveille described the experiment in Brazil in great detail.

[57] Joseph Bertrand, *Éloge historique de Léon Foucault.* Paris: Institut de France, 1882, p. 24.

[58] With the advent of Einstein's general theory of relativity early in the twentieth century, this observation had to be modified. We now know that a gyroscope located in the vicinity of a massive object will experience a *precession* of its axis of rotation. This means that the axis of rotation of the gyroscope will gyrate rather than remain constant in its direction in space. According to general relativity, two factors will produce this precession. The first is called the "geodetic effect" and is due to the fact that a massive object produces a curvature of space-time around it. The second factor is called the Lense-Thirring effect, also known as "frame-dragging." This phenomenon is due to the rotation of the massive object. As it rotates, a massive object, such as the Sun, will "drag" any reference frame we use in our calculations along with it. The phenomenon is similar to a metal ball rotating inside a container of syrup or liquid chocolate at a factory. When you look at it, you will see that as the ball is rotating, it is dragging with it the syrup that touches it. According to Einstein (and J. Lense and H. Thirring, who discovered this theoretical effect in 1918), space-time itself is dragged somewhat along with a rotating massive object. Both the geodetic effect and frame-dragging have a very small influence on a gyroscope on Earth, however.

[59] It should be noted that modern, twenty-first century compasses work on the principle of interference of laser light. Rotational motion will change the frequency of light traveling through two separate arms, meeting at a point. This shift in frequency of light in one arm relative to the other produces a measurable interference pattern, which is then electronically displayed as a change in direction.

[60] A good way to see these fashions is to look at paintings of Manet, Monet, and Degas, who worked during this period and liked to paint everyday scenes in France, including fashionably dressed people.

[61] *Journal des Débats,* December 3, 1851, front page.

[62] Jasper Ridley reports in his book *Napoléon III and Eugénie* that Louis-Napoléon had advised his son, after his own defeat, never to attempt to retake control of France through a coup d'état, saying that it had always been a "cannonball dragging around his feet" (Ridley, 1979, p. 593).

[63] J. M. Chapman and Brian Chapman, *The Life and Times of Baron Haussmann.* London: Weidenfeld and Nicolson, 1957, p. 1.

[64] Louis-Napoléon Bonaparte, *Les idées Napoléoniennes,* 1839, pp.12–15.

[65] Académie des Sciences, *Comptes rendus hebdomadaires des seances de l'Académie des Sciences.* Paris, Oct. 3, 1853, pp. 517–518.

[66] Manuscript 3710, The Letters of Urbain Le Verrier, Institut de France, Paris.

[67] Manuscript 3710, The Letters of Urbain Le Verrier, Institut de France, Paris.

[68] This letter, signed by Ildefonse Favé and written on the Imperial Cabinet stationery, is privately owned and was exhibited at *Léon Foucault, le Miroir et le Pendule: An Exhibition at the Paris Observatory,* October 16–December 15, 2002.

[69] Manuscript 3710, *The Letters of Urbain Le Verrier,* Institut de France, Paris.

[70] Most professional telescopes today are reflecting rather than refracting telescopes; some are Newtonians. Some telescopes are called Schmidt-Cassegrain, and others Maksutov-Cassegrain.

[71] Stéphane Deligeorges, *Foucault et ses pendules.* Paris: Éditions Carré, 1990, p. 110. Author's translation.

[72] Steven Weinberg, *Gravitation and Cosmology: Principles and Applications of the General Theory of Relativity.* New York: Wiley, 1972, p. 87.

[73] Biographical Dossier: Léon Foucault, Academy of Sciences, Paris. Author's translation.

[74] Biographical Dossier: Léon Foucault, Academy of Sciences, Paris. Author's translation.

[75] Stéphane Deligeorges, *Foucault et ses pendules*. Paris: Éditions Carré, 1990, p. 85.

[76] From the Archives of the Academy of Sciences, Paris.

[77] Pierre de La Gorce, *Histoire du Second Empire,* seven volumes, Paris, 1894–1905.

[78] Foucault's close friend Lissajous later wrote a moving eulogy for Foucault, which he presented to the Academy of Sciences. Much of the information in this section on Foucault's final days is based on Lissajous's account.

[79] Jules-Antoine Lissajous, *Notice historique sur la vie et les travaux de Léon Foucault (de l'Institut).* Clichy: Imp. De P. Dupont, reprinted 1875. Author's translation.

[80] Camille Flammarion, *Notice scientifique sur le pendule du Panthéon: Experience reprise en 1902.* Paris: Sociéte Astronomique de France, 1902, p. 3. Author's translation.

[81] J.G. Hagen, S.J., *La rotation de la Terre: Ses preuves mechaniques anciennes et nouvelles.* Specola Astronomica Vaticana I. Rome: Tipografia Poliglotta Vaticana, 1911.

[82] The second proof of Foucault's sine law is adapted from C. Kittel, et al., *The Berkeley Physics Course: Volume 1, Mechanics,* Second edition. New York: McGraw-Hill, 1973, pp. 114–115.

BIBLIOGRAPHY

———◆———

Académie des Sciences. *Comptes rendus hebdomadaires des séances de l'Académie des Sciences.* Paris: Institut de France. Various issues, 1851–1870.

Acloque, Paul. *Histoire des experiences pour la mise en evidence du mouvement de la Terre.* Paris: Cahiers d'Histoire et de Philosophie des Sciences, No. 4, 1982.

Acloque, Paul. *Oscillations et stabilité selon Foucault: critique historique et expérimentale.* Paris: CNRS, 1981.

Agulhon, Maurice. *Nouvelle histoire de la France contemporaine: 1848 ou l'apprentissage de la République.* Paris: Éditions du Seuil, 1992.

Aronson, Theo. *The Fall of the Third Napoleon.* New York: Bobbs-Merrill, 1970.

Balibar, F. *Galilée, Newton lus par Einstein.* Paris: Presses Universitaires de France, 1981.

Beghin, H. *Les preuves de la rotation de la Terre.* Paris: Palais de la
Découverte, 1986.

Bell, E. T. *Men of Mathematics: The Lives and Achievements of the
Great Mathematicians from Zeno to Poincaré.* New York:
Simon & Schuster, 1965.

Bertrand, Joseph. *Éloge historique de Léon Foucault.* Paris: Institut de France, 1882.

Bertrand, Joseph. *Des progrès de la méchanique: Léon Foucault.*
Paris: Revue des Deux Mondes, 1864.

Bierman, John. *Napoléon III and His Carnival Empire.* New
York: St. Martin's Press, 1988.

Bonaparte, Louis-Napoléon. *Les idées Napoléoniennes.* Paris: 1839.

Boyer, Carl B. *A History of Mathematics.* New York: Wiley, 1968.

Boyle, David. *Impressionist Art.* Vancouver: Raincoast, 2001.

Bresler, Fenton. *Napoleon III: A Life.* New York: Carroll & Graf,
1999.

Chapman, J. M. and Chapman, Brian. *The Life and Times of
Baron Haussmann.* London: Weidenfeld and Nicolson, 1957.

Ciufolini, Ignazio, and Wheeler, John A. *Gravitation and Inertia.*
Princeton, NJ: Princeton University Press, 1995.

Deligeorges, Stéphane. *Foucault et ses pendules.* Paris: Éditions
Carré, 1990.

Deligeorges, Stéphane. *The Foucault Pendulum in the Panthéon.*
Paris: Musée du Conservatoire National des Arts et Métiers,
1997.

Denizot, A. *Das Foucaultische Pendel und die Theorie der relativen Bewegung.* Berlin: B. G. Teubner, 1913.

Donné, Alfred, and Foucault, Léon. *Cours de microscopie.* Paris: Baillère, 1845.

Eco, Umberto. *Foucault's Pendulum.* New York: Harcourt, 1988.

Flammarion, Camille. *Notice scientifique sur le pendule du Panthéon: Experience reprise en 1902.* Paris: Société Astronomique de France, 1902.

Foiret, J., Jacomy, B., and Payen, J. *Le pendule de Foucault au Musée des Arts et Métiers.* Paris: Conservatoire National des Arts et Métiers, 1990.

Foucault, Léon. *Mesure de la vitesse de la lumière.* Paris: Armand Colin, 1913.

Foucault, Léon. *Notice sur les travaux de M. Léon Foucault.* Paris: Mallet-Bachelier, 1963.

Foucault, Léon. *Recueil des travaux scientifiques.* Paris: Gauthier-Villars, 1878.

Foucault, Léon. *Sur les vitesses relatives de la lumière dans l'air et dans l'eau.* Paris: Bachelier, 1853.

French, A. P. *Newtonian Mechanics: The M.I.T. Introductory Series.* New York: Norton, 1971.

Gand, E. *Application du gnomon au gyroscope ou demonstration physique du mouvement annuel de la Terre dans l'espace.* Amiens: Caron, 1853.

Gapaillard, Jacques. *Et pourtant elle tourne! Le mouvement de la Terre.* Paris: Éditions du Seuil, 1993.

Gapaillard, Jacques. *Le mouvement de la Terre: La détection de sa rotation par la chute des corps.* Paris: Cahiers d'Histoire et de Philosophie des Sciences, No. 25, 1988.

Gariel, C. M. *Léon Foucault.* Paris: Annuaire Dehérain, 1869.

Gerard, Alice. *Le Second Empire.* Paris: Presses Universitaires de France, 1973.

Gilbert, Philippe-Louis. *Léon Foucault, sa vie et son oeuvre scientifique.* Brussels: A. Vromant, 1879.

Gondhalekar, P. *The Grip of Gravity: The Quest to Understand the Laws of Motion and Gravitation.* New York: Cambridge University Press, 2001.

Grattan-Guinness, Ivor. *The Norton History of the Mathematical Sciences.* New York: Norton, 1998.

Hagen, J. G., S.J. *La rotation de la Terre: Ses preuves mechaniques anciennes et nouvelles.* Specola Astronomica Vaticana I. Rome: Tipografia Poliglotta Vaticana, 1911.

Hobson, Eric. *The Age of Capital: 1848–1875.* New York: Random House, 1975.

Holton, Gerald. *Thematic Origins of Scientific Thought: Kepler to Einstein.* Cambridge, MA: Harvard University Press, 1973.

Hoskin, Michael. *The Cambridge Illustrated History of Astronomy.* New York: Cambridge University Press, 1997.

Jerrold, B. *The Life of Napoleon III.* London, 1874–82.

Kittel, C., et al. *The Berkeley Physics Course: Volume 1, Mechanics, Second Edition.* New York: McGraw-Hill, 1973.

Koyré, A. *Chute des corps et mouvement de la Terre, de Kepler à Newton.* Paris: Vrin, 1973.

La Gorce, Pierre de. *Histoire du Second Empire,* seven volumes. Paris, 1894–1905.

Laplace, Pierre-Simon. *L'exposition du système du monde*, 1796. Reprint, Paris: Fayard, 1984.

Lissajous, Jules-Antoine. *Notice historique sur la vie et les travaux de Léon Foucault (de l'Institut)*. Paris: P. Dupont, 1875.

Mach, Ernst. *The Science of Mechanics*. La Salle, IL: Open Court, 1960.

Maitte, B. *La lumière*. Paris: Editions du Seuil, 1981.

Morin, A. *Discours prononcé aux funérailles de M. Foucault*. Paris: Firmin-Didot, 1868.

Pasachoff, Jay. *Astronomy: From the Earth to the Universe*. Orlando, FL: Saunders, 1998.

Plana, Jean. *Note sur l'expérience communiquée par M. Léon Foucault*. Paris: Académie des Sciences, 1851.

Plessis, Alain. *Nouvelle histoire de la France contemporaine: De la fête impériale au mur des fédérés 1852–1871,* Vol. 9. Paris: Éditions du Seuil, 1979.

Reda, Jacques. *Le méridien de Paris*. Paris: Fata Morgana, 1997.

Ridley, Jasper. *Napoleon III and Eugénie*. New York: Viking, 1979.

Sarda, François. *Les Arago: François et les autres*. Paris: Tallandier, 2002.

Tardieu, J. E. *Explication des phénomènes de rotation et d'orientation du gyroscope de M. Foucault*. Paris: Mallet-Bachelier, 1855.

Thomson, David. *France: Empire and Republic, 1850–1940*. New York: Harper & Row, 1968.

de Waard, Cornelius, ed. *Correspondance du P. Marin Mersenne, religieux minime.* Paris: Éditions du Centre National de la Recherche Scientifique, 1969.

Weinberg, Steven. *Gravitation and Cosmology: Principles and Applications of the General Theory of Relativity.* New York: Wiley, 1972.

INDEX